T0190646

Hot Science is a series explor... ...
and technology. With topics from big data to rewilding,
dark matter to gene editing, these are books for popular
science readers who like to go that little bit deeper ...

AVAILABLE NOW AND COMING SOON:

Destination Mars:
The Story of Our Quest to Conquer
the Red Planet

Big Data:
How the Information Revolution
is Transforming Our Lives

Gravitational Waves:
How Einstein's Spacetime Ripples Reveal
the Secrets of the Universe

The Graphene Revolution:
The Weird Science of the Ultrathin

CERN and the Higgs Boson:
The Global Quest for the Building
Blocks of Reality

Cosmic Impact:
Understanding the Threat to Earth from
Asteroids and Comets

Artificial Intelligence:
Modern Magic or Dangerous Future?

Astrobiology:
The Search for Life Elsewhere in
the Universe

Dark Matter & Dark Energy:
The Hidden 95% of the Universe

Outbreaks & Epidemics:
Battling Infection From Measles to
Coronavirus

Rewilding:
The Radical New Science of Ecological
Recovery

Hacking the Code of Life:
How Gene Editing Will Rewrite Our Futures

Origins of the Universe:
The Cosmic Microwave Background
and the Search for Quantum Gravity

Behavioural Economics:
Psychology, Neuroscience,
and the Human Side of Economics

Quantum Computing:
The Transformative Technology of the
Qubit Revolution

The Space Business:
From Hotels in Orbit to Mining
the Moon – How Private Enterprise is
Transforming Space

Game Theory:
Understanding the Mathematics of Life

Hothouse Earth:
An Inhabitant's Guide

Nuclear Fusion:
The Race to Build
a Mini-Sun on Earth

The Science of Music:
How Technology has
Shaped the Evolution of
an Artform

Biomimetics:
How Lessons from Nature can
Transform Technology

Hot Science series editor: Brian Clegg

CONSCIOUSNESS

How Our Brains Turn
Matter into Meaning

JOHN PARRINGTON

ICON

Published in the UK and USA in 2024 by
Icon Books Ltd, Omnibus Business Centre,
39–41 North Road, London N7 9DP
email: info@iconbooks.com
www.iconbooks.com

ISBN: 978-183773-078-0
eBook: 978-183773-079-7

Typeset by SJmagic DESIGN SERVICES, India.

Printed and bound in the UK.

ABOUT THE AUTHOR

John Parrington is an Associate Professor in Molecular and Cellular Pharmacology and a Tutorial Fellow in Medicine at the University of Oxford. He is the author of three previous books and over 110 peer-reviewed articles. His research focuses on how chemical signals regulate important processes in the body.

CONTENTS

WHAT IS CONSCIOUSNESS? 1

What is the material basis of the thoughts that occur inside our heads? What makes my thoughts different from yours and why do different people have distinctive personalities? Where do imaginative, creative or spiritual thoughts come from – can these really simply be the product of nerve impulses in the brain? Is human consciousness so different from that of other species or is our uniqueness more superficial than we might imagine?

These are fundamental questions – some would say among the biggest unresolved questions in science. Indeed, such is their nature that some influential philosophers doubt whether science is even capable of answering them. Because of this, it may seem presumptuous to even try and begin to provide answers to such questions in a single book, but that is what I'm aiming to do here. In so doing, I'll be drawing not only on the latest evidence from neuroscience and psychology, but also on a range of philosophical insights.

Certainly, questions like the ones above have taxed the minds of philosophers for millennia and have probably been a source of debate long before human beings first discovered ways to record our thoughts and ideas. In recorded history, one of the first people to speculate about the nature of consciousness was the philosopher Aristotle.

He believed that consciousness exists as a continuum of different types of 'souls'[1]: thus plants have a vegetative or nutritive soul, which controls their growth, nutrition and reproduction; animals have such characteristics too, but also a sensitive soul, which allows them to perceive things and move about, and they also have fears and desires; and finally humans have all of these characteristics, but also a rational soul that allows them to reason and reflect. It was an interesting viewpoint that could have been explored further scientifically, but for the next 2,000 years, such was the stifling power of religion that there were many barriers to developing a scientific understanding of consciousness.

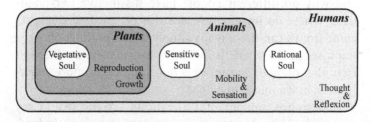

Figure 1. Aristotle's three types of soul.

Dualistic view

But things particularly began to change about 400 years ago. Inspired by William Harvey's demonstration that the heart functions like a pump, the philosopher René Descartes, one of the foremost thinkers in the world in the first half of the 17th century, proposed that the body could be seen as acting like a machine. Descartes also saw our ability to view ourselves and the world around us in a rational way as proof that consciousness was real – the basis of his famous statement 'I think, therefore I am' – and he even suggested that some aspects of human behaviour, such as unconscious reflexes, could be explained by material forces.[2] Yet he argued that the 'soul' would always remain unknowable to science. Descartes' caution on this matter may have been influenced by his religious views, and also by how little was known about the mechanistic basis of consciousness at this time. However, this 'Cartesian dualism' relating to differences between body and mind has been a problematic feature of discussions about consciousness ever since.[3]

In the late 17th and early 18th centuries, other philosophers became more willing to subject consciousness in all its aspects to scientific enquiry. John Locke and then David Hume argued that the mind could be viewed as a 'blank slate' and each individual human consciousness therefore was just the accumulation of experiences acquired since birth.

Although a properly materialist view of consciousness – meaning an explanation of the human mind without recourse to supernatural causes – a problem with this understanding was that it did not explain how each individual mind feels like a unified, individual phenomenon, rather than just a mass of unconnected experiences.[4] In fact, we will see later that this 'binding problem' has become a major issue of debate and study in modern neuroscience. Another problem with Locke and Hume's view of the mind is that by ignoring the role of differences in individual biology in the formation of consciousness, they did not explain why two different people growing up in a very similar environment can turn out radically different in their abilities, personality, temperament and so on. This viewpoint also failed to explain why human consciousness seems so different from that of animals.

One philosopher of this period who did engage with the question of how human consciousness might be related to our biology was Gottfried Wilhelm Leibniz.[5] He suggested that the inner workings of the mind might be likened to the different pieces of machinery in a textile mill – a striking new phenomenon in the 18th century. Yet he claimed that even if one could explore in detail the insides of such a mind, while this could reveal its components, it was unlikely to take us any closer to understanding the human mind, partly because of the deeply subjective nature of each individual human consciousness, but also because of the complexity of the interactions between the mind's individual parts.

Hard problem

Leibniz's scepticism about the possibility of a truly materialist explanation of consciousness might be seen as justified given how little was known about the brain at that time. Yet while scientific knowledge about this organ has advanced dramatically, it is not clear that we are any closer to a proper understanding of consciousness. The philosopher David Chalmers has expressed this conundrum by what he calls the 'hard problem of consciousness'.[6] Chalmers believes that neuroscience may soon allow us to understand how we learn, store memories, perceive things, react instantly to a painful stimulus or hear our name spoken across a room at a noisy party. Indeed, he thinks these may be relatively easy aspects of consciousness to decipher, at least with sufficient research time and money.

In contrast, Chalmers sees the 'hard problem' of consciousness as explaining that subjective sense we have as individuals of being us, with all that implies in terms of our specific responses to, say, a sunset or a work of art, the particular way we felt when we first fell in love or any personal experience, in purely material terms. It is the difficulty in explaining such subjectivity that has led some to view the problem of explaining consciousness in this way as ultimately impossible. However, surely this viewpoint does not take us much further than we got with Descartes, at least if our aim is a materialistic view of consciousness. Indeed, some critics have accused Chalmers and other proponents of the 'hard problem' of clinging to the idea of a 'soul' that is forever unknowable to scientific methods, just like that of Descartes.[7] My personal feeling is that Chalmers has identified a real problem for explanations of consciousness that seek to describe it in purely material terms, yet I do not believe it is an insurmountable problem for science, for reasons I will outline later.

The debate about whether we will ever understand the material nature of consciousness is not just one between philosophers on one side and neuroscientists on the other. Daniel Dennett is a philosopher, but one who has also championed a very materialist view of consciousness. Dennett is a critic of idealist models of human consciousness, meaning ones not based on the material properties of our brains.[8] Such models ultimately rely on there being some kind of homunculus – meaning 'little human' – directing things from inside the brain, but this only begs the question of who controls the brain of the homunculus and so on. Instead, Dennett proposes a 'bottom-up' approach, which sees the mind as the combined product of unconscious, evolved processes that somehow combine to provide the appearance of an individual 'I', yet in reality has no conscious entity at its core or, for that matter, a specific place in the brain where 'it all comes together'.

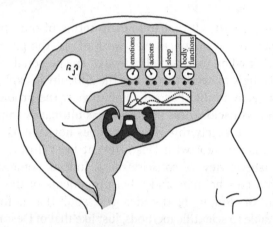

Figure 2. Cartesian dualism and the idea of a homunculus controlling the brain.

Dennett has also criticised theories of consciousness that rely on what he calls 'skyhooks' – explanations of complexity that do not build on lower, simpler layers. Yet ironically, in his own view of consciousness, Dennett has used what I consider a skyhook: memes.[9] Nowadays, a meme tends to signify those images or video clips – often humorous, cringeworthy or carrying some life message – that can spread so rapidly on social media. The term was first used in 1976 by Richard Dawkins, in his book *The Selfish Gene*, to describe an idea, behaviour or style that spreads between people – examples being 'tunes, ideas, catch-phrases, clothes fashions, ways of making pots or of building arches'. Dawkins argued that 'just as genes propagate themselves in the gene pool by leaping from body to body via sperm or eggs, so memes propagate themselves in the meme pool by leaping from brain to brain via a process which, in the broad sense, can be called imitation'.[10]

As a description of how popular images can spread rapidly across the internet, almost like a viral infection, this was a remarkably prophetic and insightful vision. However, using memes as an example of how human consciousness works seems to me to be an idealist viewpoint because it suggests that ideas are independent entities with a separate identity from the minds they inhabit, when what we really need to do to establish a materialist theory of consciousness is to show how ideas originate organically within individual human brains, as well as passing between them.

Here, Chalmers' point about there being a 'hard' problem within consciousness also seems relevant. For another criticism that could be made of Dennett's view is that, even if his 'bottom-up' approach that sees consciousness as something emerging from a mass of unconscious neural impulses is true, that still leaves the problem of explaining the very subjective nature of an individual human consciousness in material terms. Even if we also accept the idea that consciousness is the product of many unconscious processes distributed across the brain, that still leaves the question of why it is that as human individuals we have that very clear and vivid sense of ourselves as individual entities.

One area of potential confusion when discussing consciousness is what we mean by this term. We saw how Aristotle believed there were three types of consciousness, with only humans possessing the rational, 'higher' form. The idea that humans are unique compared to other species continued with the Judeo-Christian philosophical tradition. Yet the past half millennium has seen an erosion of the idea that there is anything unique about human beings and our place in the universe. This trend began with Nicolaus Copernicus' demonstration in 1543 that, instead of being at the centre of the universe, the Earth is merely a satellite of the Sun, which we now know to be just one star among many others.

Mental spectrum

A further blow to our egos came with Charles Darwin and Alfred Russel Wallace's theory of evolution by natural selection, most famously expounded in Darwin's *The Origin of Species*, published in 1859, which showed that humans are only one of many branches on the tree of evolutionary change.[11] We now appreciate that while life itself has existed on Earth for 4 billion years, humans only diverged from our closest relative, the chimpanzee, between 6 to 10 million years ago. Moreover, comparisons of the human and chimp genome show these are 96 per cent similar in DNA sequence. Because of this genetic continuity between humans and other species, some philosophers and neuroscientists have begun to look for such continuity in consciousness. For instance, neuroscientist Christof Koch has recently claimed that 'consciousness is ... probably present in most of metazoa, most animals, [and] it may even be present in very simple systems like a bacterium'.[12]

The idea that consciousness is something shared by many species underlies a now famous article by the philosopher Thomas Nagel in which he asked: 'What is it like to be a bat?'[13] In this, Nagel makes two major assumptions. One is that 'conscious experience is a widespread phenomenon' present in many animals, particularly mammals. Another is that such an experience has a 'subjective' character. Following this, Nagel argues that since bats have a different sensory apparatus to humans, relying on sonar to navigate the world far more than by a visual system like most people, it would be very difficult to imagine what it would be like to be a bat. Nagel uses this fact to question whether scientific methods can ever reveal the true nature of consciousness in a materialist, objective fashion, or whether this will always be outside science's reach. His argument is similar to Chalmers' notion of a 'hard problem' in consciousness, but with added emphasis, since imagining what it feels like to be a bat is surely even more difficult than imagining being another human.

However, what if Nagel's assumptions are incorrect? What if the self-conscious awareness that we humans generally mean when we talk about having a consciousness is not shared with other species? What if, as a consequence, there is also no sense that any other species can have a subjective sense of themselves as an individual, but rather they are a complex mass of feelings and sensations but with nothing like the individual identity we take for granted as individual humans? So while other species may be conscious in one sense, this is very different from our human self-conscious awareness. This may seem a bold claim, but it explains one major difference between humans and other species – our capacity for transforming the world around us with each new generation; it is this that in 40,000 years has allowed us to go from scratching a living from the Earth to sending rockets to Mars.

In contrast, no other species on the planet, including our closest biological cousins, the great apes, has shown any capacity to transform the world around them in the way that human beings do. Not that this capacity is always a good thing, as witnessed by the fact that human civilisation may be heading for catastrophe in the form of global warming, chemical pollution, mass extinction of other species or the threat of a nuclear holocaust, but it is a unique capacity nevertheless, and I would argue that it is a direct manifestation of something else distinctive about human beings, namely our powers of conceptual thought and language, coupled with our ability to design, and redesign, new types of tools and technologies. So let us now look at how these capacities arose during our evolution from apes and how this led to human consciousness becoming a very specific entity on Earth.

TOOLS AND SYMBOLS 2

An idea I will keep returning to in this book is that human self-conscious awareness arose as a consequence of two other unique human attributes – our capacity for language and our ability to continually transform the world around us by designing and using tools. However, there is also another vital factor in what makes humans unique, which is our brains: these are not just much bigger than those of other primates, but radically different in structure and function. I see these three capacities as interconnected in terms of their evolution.

Is it really true that human language and the way we use technologies to transform the world around us are unique to our species? Some would dispute this uniqueness. For instance, they might point to the fact that other species communicate through various sounds and gestures or to evidence that other primates and even some other types of animal, such as crows, develop and use tools. I believe that such arguments miss the key qualitative difference between human beings and other species. To understand why, it is worth looking in more detail at how humans first evolved from apes.

Surprisingly, the first person to identify the correct sequence of human evolution was not Charles Darwin, as one might expect, but Friedrich Engels.[1] Despite being known primarily as a political activist and thinker, Engels also had a profound interest in natural science, and in an essay he wrote in 1876, he proposed that humans first began to diverge significantly from other primates when our ancestors started walking on two legs. This freed the hands for using and designing tools, and as a consequence, proto-humans began using tools in a systematic way to transform the world around them. Importantly, such design and subsequent use of different types of tools was carried out with other proto-humans in a socially cooperative manner. Because of the need to communicate with their neighbours about how to carry out such innovative actions, our ancestors also began to develop the first forms of language. Subsequently, the development of both systematic tool design and use, and language, led to a dramatic growth and restructuring of the brain.

Engels' proposed sequence of events differed from that of Darwin, who argued that the development of a large brain preceded bipedalism, tool use and language.[2] However, subsequent accumulated fossil and DNA evidence have confirmed that what Engels proposed is correct, although, in contrast to the linear progression that he envisaged, human evolution has been more a case of multiple proto-humans with different characteristic features co-existing and with many blind alleys ending in extinction along the path to *Homo sapiens*.[3]

If some other species use tools, what is distinctive about human tool use? A key difference is that tool use by other species tends to be both occasional and also very limited in the type of tools that are created. In contrast, a unique feature of our species is that practically all of our interactions with the world are through tools that we have created.

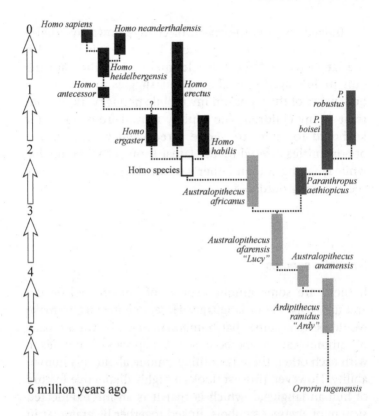

Figure 3. Evolution of humans from apes.

As well as the fact that tool use is systematic to the way that humans interact with the world, in contrast to the more accidental and occasional use of tools by some other species, another key feature of human beings is the way we are continually in a process of inventing new types of tools and technologies. While this was a relatively slow business early in our prehistory, in modern times, it has accelerated to the point that we now take for granted the way that novel technologies rapidly develop during the lives of each new generation.

Indeed, such is the speed of change nowadays that a characteristic feature of modern life is keeping up with the latest technological developments. I doubt I am the only middle-aged individual who struggles to understand how many of the 'apps' on my mobile phone work, in contrast to my children who rapidly moved from computer keyboard to tablet to phone screen and who seize the opportunities offered by the latest forms of social media, while laughing at their father's continuing use of ones they see as totally outdated.

Symbolic species

If these are some unique aspects of human tool design and use, what about language? Here, one mistake to make would be to assume that human language is just a means of communication, and because other species communicate with each other, there is nothing unique about this human ability. However, this overlooks a highly distinctive feature of human language, which is that it is an interconnected system of abstract symbols, linked together by grammar in such a way that it can convey complex meaning. It is for this reason that only human beings are able to use language to convey complex ideas like past, present and future, individual versus society, location in space and even more abstract concepts.

Showing how this capacity is a unique feature of human biology, intensive attempts to teach sign language to our closest primate cousins, chimpanzees and gorillas, in the 1970s, demonstrated that, while such species can learn to associate words with objects and even emotions, these other primates lack our grammatical capacity and as a consequence the ability of human beings to represent the world conceptually through abstract symbols.[4] This is a key difference between humans and apes that I will explore in more detail later.

Underpinning the unique human attributes of systematic tool use and language is a brain that has evolved to give us the capacity to handle these attributes in a meaningful way. Indeed, we will see later that there is a good reason for this, for both our use of tools to transform the world around us, and our use of language to communicate with each other, have, during our evolution and also in the development of each individual human being, led to the transformation of the human brain in a highly distinctive way. A useful analogy here is to see tool use and language as activities that do not just guide our interactions with the external world, but also act as 'mental tools' that have transformed the brain in the process.

If systematic use and development of technology and language capacity are unique features of our species, what about the way our brain works compared to other species? Here we face a problem for while it is possible to objectively study tool use and language in humans and compare this to the abilities of other animals, what goes on inside an individual brain of a human or other species, particularly at the subjective level, is far harder to assess.

Yet I believe we can gain an objective and scientific understanding of what makes human consciousness unique, both through psychological analysis and by studying the ways in which the human brain differs from those of other species in both structure and function. I will also be exploring what we have learned so far in this quest later in this book. However, what we also need is a firm conceptual foundation for such analysis and in particular one that requires a better understanding of the links between human thought and language, and how language has transformed human thought in a way that makes it qualitatively different from the thoughts of other species. These are topics that I will look at in this and later chapters.

For some, the idea of using scientific methods to understand the basis of an individual person's thought processes can seem like an impossible task. This surely is at the heart of David Chalmers' 'hard problem of consciousness', meaning the difficulty of explaining human consciousness in objective, scientific terms because of the impossibility of 'getting inside' the head of another person and observing their thoughts. In fact, we even face a problem in trying to get inside our own heads, since the moment we try to explain our innermost thoughts, we are potentially altering them in that process of explanation.

Inner speech

One of the early pioneers of psychology, William James, likened the problem of attempting to observe our own thoughts to 'trying to turn up the gas quickly enough to see how the darkness looks'.[5] Yet I believe we can gain important insights into both our own and other people's thought processes because of what I have said previously about two unique aspects of human beings compared to other species, these being our ability to design and use tools and our capacity for language. I would also argue that our language capacity in particular shapes our thought processes in ways that we can begin to study and understand.

Someone who played a pioneering role in this respect was the psychologist Lev Vygotsky. He argued that 'egocentric speech' – children's tendency to talk to themselves as they play – is the first stage in a child starting to organise their actions using words and that while this form of speech seems to disappear, what is actually happening is it becomes internalised as 'inner speech', which in adults plays a central role in the organisation and development of our thoughts.[6] So by studying egocentric speech it should be possible to gain insights into the particular character of inner speech. It is also possible to understand inner speech better by studying outer speech and then extrapolating to the likely character of the inner speech from which the outer version originates in the brain. Finally, we can introspectively analyse our own inner speech and thereby try and learn about its character.

Such studies combined suggest that inner speech differs from outer speech in some important ways.[7] It is likely to be much more rapid, and far more fluid in meaning, than the speech we use in conversation with others. There are also probably different types of inner speech, ranging from that which emerges from our innermost, half-formed thoughts to the type that structures our outer speech when we express ourselves to others.

An important consequence of human inner consciousness being structured by language is that this gives a particular social dimension to human consciousness that is lacking in other species. Originating as they do in society, words necessarily infuse our thoughts with social meaning. Given that our inner speech is not merely a reflection of our present circumstances, but also carries with it the memory of past ones, this means that our inner consciousness must be deeply infused with past social interactions that we have had with other people, for instance, parents, siblings, teachers, friends and colleagues.[8]

While language is the primary structurer of inner human consciousness, we should not forget what I have said about tools also being central to what it means to be human. For a distinctive feature of human beings is not only that we interact with the world around us through tools, but that these continually change with each new generation. This is likely to significantly affect not just humanity as a whole but each individual consciousness.

Cultural tools

One has only to think of the impact that the invention of reading and writing must have had on human thought processes, or more recently the ways we have begun to communicate using the internet, to see how profoundly such changes may affect the workings of the inner human psyche. Particularly relevant to the ways that social media may now be affecting individual human consciousnesses is that this is not only based around the written word, but also visual images, videos, music, abstract symbols and so on. In fact, not only spoken language but a variety of what we might call 'cultural tools', which can include music, visual art, literature and mathematical and scientific symbols, are all distinctive to humanity and could each have an influence on human consciousness.[9]

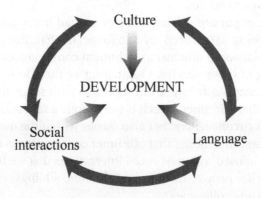

Figure 4. Vygotsky's theory of cognitive development.

If language and also other cultural tools are central to structuring our inner consciousness, where does this leave thought? Here it is worth returning to what I said previously about the similarities and differences between humans and other species. Given that human brains have much in common with those of other mammals, particularly other primates, not just at basic molecular and cellular levels but also in terms of different brain regions and their interconnections, it seems likely that at the most basic level of thought processes we also share much in common. Yet the lack of language and other cultural tools in other species surely must have a profound impact on their inner consciousness.

I said previously that a key feature of language that distinguishes it from animal communication is that it is an interconnected system of abstract symbols linked by grammar. However, since only we have such a system, surely this means that the inner consciousness of other species cannot have that language-based, conceptual underpinning that we take for granted. In fact, later in this book I will be looking at the question of whether non-language-based types of conceptual thought might exist in other species, particularly non-human primates. None of this is to say that other species cannot have very sophisticated behaviours, feelings and social interactions, but this surely raises the question of whether any other species has a unified sense of self and their place in the world as we do.

To return to something I mentioned previously, Thomas Nagel's essay 'What is it like to be a bat?', although Nagel intended this question to reveal the limits to how much we can ever know scientifically about the subjective character of consciousness, for me this question is itself problematic. This is because not only could we never know what it feels like to be a bat, maybe neither could a bat, at least not in terms of being able to express or be self-consciously aware of such a feeling. The key then to understanding human consciousness is language and tool-use, but that still leaves the question of how that consciousness manifests itself in terms of molecular and cellular signals in the brain, activities in different brain regions and interconnections between those regions, which are issues I will now begin to explore in more detail by looking at nerves and brains.

NERVES AND BRAINS 3

If we want to understand how human consciousness emerges from a material object, the brain, ultimately we need to know more about how the brain functions as one of the body's organs. Here though we face a problem. For although our understanding of how other organs function is now fairly clear, this is still far from the case when it comes to the brain. I mentioned previously how William Harvey transformed our view of how the human body works when he demonstrated that the heart can be thought of as a pump.[1] We also now understand how the liver stores and creates energy and destroys toxins, the kidney flushes out waste products from the blood and maintains its concentration of salts and other substances vital for life, and the stomach and intestines digest and absorb food.

Of course, every year we gain important new insights into how our organs work.[2] A strength of science is its ability to cast new light on processes thought to be already fully understood. Yet, in general, our understanding of how most human organs function is well advanced. In contrast, although our understanding of the fine detail of the molecular and cellular processes that occur in the brain have increased dramatically over recent decades, I believe that even most neuroscientists would agree that things are very different when it comes to understanding how the brain's components work together as a unified whole.

In order to advance our own understanding of brain function and how this relates to human consciousness, it would be useful to begin by defining the brain's basic components. The base unit of the brain is the nerve cell, or neuron.[3] In fact, there are hundreds of different types of neurons, playing specific roles in the brain that we still only incompletely understand, but what they have in common is that they are connected together in electrical circuits. Neurons have three parts – a cell body, tendrils called dendrites that receive incoming signals from other neurons and a long protuberance called an axon, that conveys signals to the next neuron in the circuit. When a neuron is stimulated through its dendrites, this triggers an electrical impulse called an 'action potential', which races along the axon at speeds that can reach around 200 mph in the fastest human neurons. The speed at which impulses can travel along an axon are greatly enhanced by a structure called a 'myelin sheath'. This is a fatty layer that both protects the axon but also allows the neural signal to jump between gaps in the sheath in a process called 'saltatory conduction'. At the end of the axon, the electrical impulse is transmitted to a dendrite on the next neuron in the circuit, at a gap between neurons called a 'synapse'. The action potential does this by stimulating the release of a chemical called a 'neurotransmitter' into the synaptic gap. The neurotransmitter diffuses across the gap and, at the other side, it acts like a key in a lock to trigger a cellular response that activates a new action potential in the target neuron.

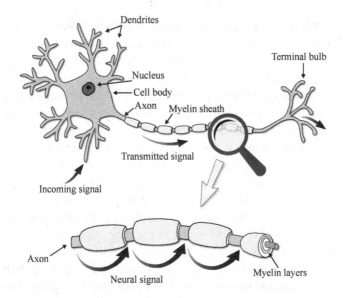

Figure 5. The nerve cell and transmission of signal.

Connectome challenge

Such is the importance of electrical circuitry for brain function that it has led to the idea that if only we could map all the neural connections in the human brain – known as the 'connectome' – then this would reveal how our brains work and, through this, consciousness.[4] Personally, I can see all sorts of potential value in knowing more about the types of neurons in the human brain and their interconnections. A central theme in biology is that structure tends to determine function and vice versa. Therefore, any new information that helps us understand the overall layout of circuitry in the brain, particularly if that includes information about the different types of neurons in that circuitry, could be very valuable.

We should not underestimate the challenge, though, of the connectome project. The human brain is estimated to contain around 86 billion neurons, linked by 100 trillion connections. As a comparison, a recent connectome project that focused on the mouse brain, which contains in total approximately 70 million neurons, managed to map only 75,000 neurons, which is only a thousandth of the neurons in the mouse brain.[5] It was a phenomenal piece of science, but it does put in perspective the difficulties in expanding to the scale required to fully map the mouse brain, let alone that of a human being.

Another problem concerned with only focusing on neurons and their connections is that these cells only make up around half of the cells in the brain. The other half is made up of glial cells.[6] For years, these were thought to have a passive role in the brain, an assumption implicit in their name, which is derived from the Greek word for glue; thus it was assumed that glial cells filled in the gaps between neurons, providing merely structural support. Yet it has recently become clear that a range of different types of glial cells play many important roles in the brain. Glial cells include oligodendrocytes, which produce the myelin sheath I mentioned previously and help to enhance conduction of electrical impulses along axons; astrocytes, which fine-tune neurons' synaptic connections and 'prune' brain regions where there is too much neuronal growth; and microglia, which act like a kind of immune system in the brain, combatting viruses and other pathogens and repairing damage.

Because glial cells are now recognised to play such important roles in the brain, any model of consciousness that only focuses on the neuronal circuitry runs the risk of ignoring half of what is going on in the brain at the cellular level. To be fair, the mouse connectome study I just mentioned included the presence and potential activity of glial cells in its consideration of what is going on in the regions of the brain that it was investigating.

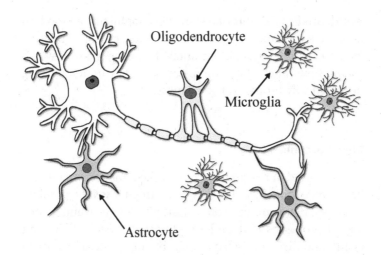

Figure 6. Neurons and neuroglial cells.

Another potential problem with most connectome studies is their focus on non-human species. There is a good reason for this. In the study mentioned above, recordings were made of the electrical activity in the neurons of the visual cortex – the part of the brain that allows us to see – of a mouse engaged in running around its cage and doing the things mice do. After hours of recording neuronal activity in this way, the mouse was killed (an unfortunate but necessary aspect of this type of research) and its brain was removed. The brain was then sliced into microscopically thin sections and the interconnections between the neurons, and also the position of glial cells in the visual cortex, were determined.

Having obtained all this data, it is hoped that it should now be possible to interrelate the electrical signals previously recorded in the visual cortex, the mouse's physical activities during the recording period and the mapped 'connectome', and so gain insights into just how that combination of electrical activity and neural connections, combined with inputs from glial cells, allow a mouse to navigate its way around the world. Studies like this one show how sophisticated modern neuroscience, aided by advances in brain imaging, microscopy and computation, is becoming. However, it is also worth asking how much such studies advance our understanding of specifically human consciousness.

Model organisms

Animals used in research are often called 'model' organisms, because the hope is that by studying them it will be possible to model, that is mimic, some of the physiological and pathological processes that underlie different aspects of human health and disease.[7] By so doing, the hope is that scientists will not only advance our understanding of basic biological mechanisms, but also devise new ways to diagnose and treat human diseases.

Indeed, such research has been fundamental in furthering our understanding of how many human organs work and in producing diagnoses and treatments for human conditions such as heart disease, diabetes and cancer. The reason is that, in general, the way that the human heart, liver, pancreas and so on function seems very similar to how organs, glands and tissues work in other mammals. Indeed, such are the similarities that exist at the basic molecular and cellular level between even quite distantly related species that major discoveries have been made about human cellular metabolism, cell division and embryo and brain development in organisms as different from humans as flies and worms.

What about human consciousness? Can animal models play a role in furthering our understanding of this phenomenon? A central claim of this book is that, despite our having evolved from apes, human consciousness has undergone a major shift, thanks to the input of systemic tool use and language, and the effect that this had upon the structure and function of the brain has radically changed the basic nature of our consciousness. So does this mean that animal models are limited in what they can tell us about this phenomenon?

One answer to this question would be to point to the basic similarities at the molecular level between human neurons and glial cells and those of other mammals. Even in terms of gross structure, there are many similarities between the human and mouse brains.[8] It is for this reason that studies in mice aimed at understanding the basis of sensory processes, like vision, hearing, touch and taste, are likely to have great relevance for furthering our understanding of these processes in humans. There are also many parallels between what we call the emotions and animal instinctual responses, as I will explore later in this book. The similarities are even greater between humans and other primates, and later I will discuss some valuable insights that have been emerging from recent studies of monkeys.

However, if we really want to understand what are sometimes called the 'higher' aspects of human consciousness, I believe there is a limit to what we can learn from animal studies alone. This is because even if our brains do have much in common with those of other mammals in terms of basic molecular and cellular structure, my prediction would be that there will be major differences, even at such basic levels, because of the way the human brain has become transformed, both during our evolution and as we develop from babies, in unique ways due to the influence of tool use and language. For this reason, I believe that studies in animals need to be complemented by an examination of what is going on in a human brain in order to understand how human consciousness emerges from such a brain.

Here, though, we face a challenge. Some people do leave their bodies to science and molecular and cellular analysis of their brains can then be carried out. Such analysis has also been performed on the brains of human foetuses that did not reach full term. Electrical recordings can also be carried out on neurons in the brains of consenting individuals having surgery to treat brain disorders, while they are asked to carry out some verbal or other sort of task. Importantly, it is also possible to gain insights into human brain function using non-invasive imaging techniques, such as functional magnetic resonance imaging (fMRI), but clearly there are limits to the sort of experiments that can be carried out on people compared to animals.[9]

Genomic insights

One important source of information for neuroscience in recent decades does not involve analysis of brains at all, but rather of genomes. When the DNA sequence of our species' genome was published in 2003, John Sulston, who led the Human Genome Project in Britain, claimed that 'we now hold in our hands the instructions to make a human being'.[10] The idea was that since the genome is often referred to as the 'blueprint' of a species, by studying the human genome sequence and comparing it to the sequences of other species, including our closest biological relatives such as chimps and gorillas and the extinct proto-human species, Neanderthals and Denisovans,[11] it should be possible to determine which genetic information is unique to human beings and to use that information to explain how we differ from other species, including our distinctive capacity for self-conscious awareness.

Some of this genome analysis is providing important insights into how our brains have evolved into such unique entities. One of the most noticeable characteristics of the human brain is its size, with our brains being three times bigger than those of chimpanzees. We are now starting to understand the genetic basis of this difference; for instance, a recent study revealed that a gene named NOTCH2NL, which is only found in humans, Neanderthals and Denisovans, may play an important role in the distinctively large size of the human brain.[12] Neurons are formed during the development of the embryonic brain from neural 'stem cells'. Like other stem cells in the body, the neural version has the ability to both divide repeatedly and also to 'differentiate' into more specialised cell types. NOTCH2NL also delays the differentiation of neural stem cells in the human brain, thereby allowing them to divide for longer and produce more neurons. Intriguingly, NOTCH2NL arose in our ancestors' genomes 3 to 4 million years ago, having developed as an offshoot of another gene named NOTCH2 that is present in many other species, which seems significant given that this is about the same time that the brains of proto-humans first began to grow in size.

The dramatic growth of the human brain has certainly been a key factor in the development of human consciousness, by increasing the number of neurons and glial cells in the brain and therefore its processing power. However, a better understanding of its genetic basis, although critical to our understanding of what it means to be human, does not in itself tell us how having a bigger brain has allowed us to develop the self-conscious awareness that, as I have argued previously, is a key and unique aspect of human consciousness. For that, we also need to know how the different parts of the human brain function and how they are connected to each other, as well as how our brain's growth affected its function. This is an issue we will now begin to examine by looking at how brains evolved and at brain structure.

EVOLVING MINDS 4

Life on Earth is based on two principles – survival and repro-duction.[1] While reproduction involves the transmission of genetic information to the next generation, survival requires the ability to stay alive and healthy in the face of what can often be a highly challenging environment. This ability is bound up with the capacity to sense changes in that environment and respond accordingly. In the simplest organisms, such as bacte-ria, this sensing of the environment involves changes through proteins on their surface called 'receptors'.[2] There are hun-dreds of types of these on a typical bacterium, each attuned to a specific chemical in the environment. If a receptor recognises a chemical that signifies food, this triggers a response that tells the bacterium to swim towards the chemical's source with the help of another protein on its surface that spins around like a tiny propellor. However, if the chemical registers as toxic, this triggers a movement away from it.

While unicellular organisms have existed on Earth for at least 3.5 billion years, large multicellular lifeforms only emerged much later, some 600 million years ago.[3] This may be because a major challenge for such multicellular organisms has been to find ways to recognise changes in the external environment but also coordinate the activities of different types of cells within the organism, what might be seen as monitoring the 'inner environment'. The need to find ways to perform both activities eventually led to the development of nervous systems.

Jellyfish are a simple type of large multicellular organism whose nervous system consists of sensory neurons, which pick up signals from the environment, 'motor' neurons, which trigger a response in the organism, and 'interneurons', which coordinate these two activities in the central nervous system (CNS).[4] In fact, these three basic types of neurons are also found in more complex organisms, including humans. In the case of a jellyfish, contact with some object in its environment that could signify a potential foodstuff then triggers the release of toxic chemicals that have evolved to allow the jellyfish to kill its prey (and may unfortunately also harm a human swimmer who has strayed into the jellyfish's path). Alternatively, the jellyfish's nervous system can stimulate a movement away from a predator, which it does by contracting powerful muscle cells around its gelatinous bell.

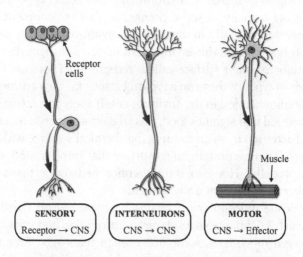

Figure 7. Three main classes of neurons.

Brain complexity

As multicellular organisms evolved to become more complex, their nervous systems needed to become more complex too, in order to coordinate the increasing number of different cell types. In particular, the interneurons multiplied and eventually became so numerous and concentrated in one part of the organism that they gave rise to the first brains.[5] As such, the human brain may be seen as the culmination of an evolutionary process that goes back hundreds of millions of years, and even further back, if we also see the sense–response systems of bacteria as a direct precursor of nerves and brains, since brain cells also have receptors on their surface. That means that our brains have much in common with the brains of other species that do not have self-conscious awareness. Yet if human consciousness is as distinctive as I claim, its unique properties must be based on equally unique features of the human brain. Therefore, in this chapter and later in this book, my goal will be to explore what seems similar to the brains of other species and what is unique in the human brain. A good place to start will be to look at the structure of our brains and how these are both similar and different from those of other species.

The human brain, like those of many other complex organisms, has three main parts: the forebrain, midbrain and hindbrain.[6] The forebrain includes the cerebrum, derived from the Latin word for brain, which has two hemispheres and a highly folded surface cortex – derived from the Latin for bark, for it does indeed wrap around the brain like bark around a tree. The cortex is particularly enlarged in humans, being in places double the area it should be for a typical primate of our size.[7] It is the brain region most thought to be associated with 'higher' brain functions, such as reasoning, planning and problem solving. Yet showing how brain regions can have multiple roles, the cortex is also involved in the regulation of movement, perception, visual processing and recognition of sounds and speech.

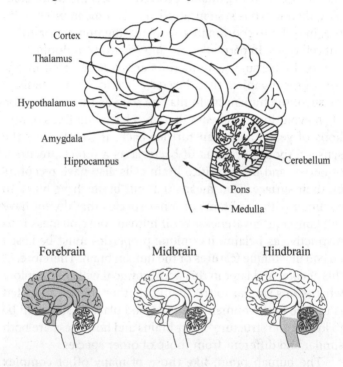

Figure 8. The different regions of the human brain.

Also buried within the forebrain are the various parts of the limbic system, which is sometimes referred to as the 'emotional brain', or even the 'lizard brain', to reflect the fact that this brain region is very ancient from an evolutionary standpoint and regulates some of the most basic life processes.[8] This region contains the thalamus, which acts somewhat like a central postal-sorting depot, channelling information into and out of the cortex. The hypothalamus – which, as its name implies, lies underneath the thalamus – regulates thirst, hunger, desire, reproduction and the body clock. The amygdala plays a central role in processing emotions, being involved in generating fearful, but also pleasurable, sensations. An adjacent region, the hippocampus, has been shown to play important roles in memory.

The midbrain is also involved in processes such as vision, hearing and movement. The hindbrain includes the pons, which controls sleep; the medulla, which regulates vital functions such as breathing and heart rate; and the cerebellum, meaning 'little brain' because, like the cerebrum, it has two hemispheres and a highly folded surface. This brain region is involved in regulating and coordinating movement, posture and balance, but recent studies have indicated that it also plays important roles in creativity and imagination.

In fact, there is an ongoing debate about the extent to which mental functions can be assigned to specific regions of the brain or whether they are more distributed across the whole brain. The idea that specific brain regions have distinct functions first became popular in the 19th century when clinicians began to identify and publicise rare individuals with specific defects in their behaviour. For instance, in 1861, the clinician Paul Broca treated a man who could not form grammatical sentences; instead, he just repeated one nonsensical word, 'tan'. When the man died, Broca dissected his brain and found damage in a region on the left side of the cortex that is now known as 'Broca's area'.[9] In contrast, fifteen years later, another clinician, Karl Wernicke, studied individuals who could produce words, but not understand them, and this defect was later linked to a left cortical region now known as 'Wernicke's area'.[10] These discoveries helped give rise to the idea that our capacity for language production and reception are localised to distinct regions in the brain.

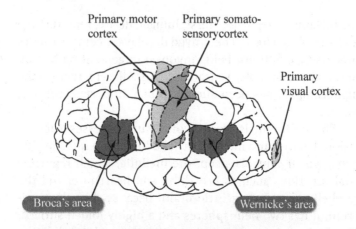

**Figure 9. Broca's and Wernicke's language
areas in the human brain.**

As I will discuss later, studies of individuals who have
suffered damage to their hippocampus showed that they
have memory defects, thereby illustrating the important
role that this brain region plays in this process. However, I
believe we have to be careful about claims that if a particular human capacity is lost when a brain region is damaged,
this then shows that this region is the sole source of this
capacity in the brain. For just as removing a spark plug from
a car and finding that the car will no longer start does not
show that the spark plug is the sole object that makes the
car function, so studies of people with lesions in particular
brain regions may help us understand the roles those regions
play in particular human capacities, but do not tell us which
other regions in the brain might also play a role.

Mental mapping

Another approach that has helped to strengthen the idea that
particular human capacities have very specific locations in

particular brain regions involves electrical stimulation of the brain of consenting individuals undergoing brain surgery for conditions such as epilepsy. During such operations, in which patients are conscious, particular regions in their brain are stimulated and then the patient is asked what feelings this invokes in them. The earliest study of this type by the neurosurgeon Wilder Penfield in 1950 led to the finding that the 'somatosensory' part of the brain cortex that receives information from the senses can be viewed like a map of the body.[11] Such a map can be drawn as a distorted homunculus with the parts of the body shown, not as their actual relative size, but as the amount of space they occupy on the cortex. A similar type of map has been determined for the body's 'motor' responses. An interesting feature of both sensory and motor maps is how much space is taken up in the cortex by nerves going from and to the lips, tongue and pharynx, and to and from the hands and digits. This fits with what I have said about the role of tool design and use, and language, in the development of human consciousness. Other studies have identified regions of the brain involved in hearing, vision, taste and smell.

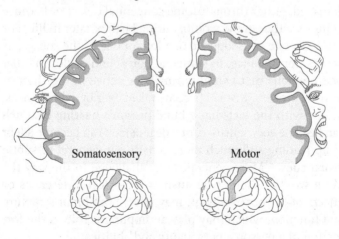

Somatosensory   Motor

Figure 10. Cortical homunculus.

While findings such as the ones mentioned above support the idea that brain functions are localised to specific regions of the brain, there is also evidence that the brain works in a much more singular fashion. One very important recent shift in our understanding has come from the discovery that brain waves of different frequencies seem to play important roles in coordinating the activities of different regions in the brain. This discovery has led the neuroscientist Earl Miller to claim that in contrast to previous viewpoints that saw the brain as 'a giant clock, and if you figure out each gear, you'll figure out the brain', the new findings suggest that the brain is better understood as 'networks interacting in a very dynamic, fluid way', with brain waves key to the coordination of such networks.[12]

Something else to make us wary of viewpoints that overly emphasise the idea of the brain as similar to a mechanical object like a clock with different gears is that recent studies have shown that the brain is far more 'plastic', that is functionally and structurally changeable, not just in children but also in adults, than had been previously thought.[13]

Plasticity can be particularly important if the brain is damaged, either through some congenital defect or because of injury either in the womb, during birth or later in life. For instance, one consequence of a loss of one of the senses, as occurs, for instance, in someone born blind or deaf, is that brain regions that would normally be active in the sight or hearing process, instead become taken over by other senses, in line with the fact that a blind person's hearing or touch can become very sensitive, or a deaf person can become adept at lip reading and much more sensitive in general to subtle visual cues.[14] More generally, any challenge to the way the brain works, whether because of biological differences or effects of the environment, may affect both brain structure and function, which may play an important role in the formation of a person's personality and abilities.

Human qualities

Perhaps one of the most interesting aspects of recent studies of the dynamics of brain structure and function are findings that indicate that the human brain differs from those of other species in the functions of its different brain regions and also their interrelationships. For instance, I mentioned earlier that the cerebellum, formerly thought only to be involved in the learning of repetitive movements of the sort that allow us to ride a bike or throw a ball into a net, has recently been shown to play a role in human creativity and imagination. A brain imaging study led by neuroscientist Alex Schlegel found that the cerebellum became active in human volunteers asked to imagine specific shapes and then mentally combine them into more complex figures or dismantle them into their separate parts.[15] This further role for the human cerebellum seems to be linked to novel connections in the brain between the cerebellum and the cortical regions of the cerebrum.

Another study has revealed that certain types of glial cell – the non-neuronal brain cells I mentioned previously – extend their projections over much greater distances in the human brain than is the case in other primates.[16] Finally, a gene called TH, which controls production of the neurochemical dopamine – whose functions I will discuss in more detail later, but is particularly associated with pleasure and reward – is switched on in some human brain regions but not in these regions in other primates.[17] All of this points to unique differences in the interconnections of the human brain, but also a transformation of the functions of some brain regions and even the roles of neurochemicals in the human brain.

To some extent such findings are mere hints, with a huge amount of work still to be done to identify just how different the human brain is in terms of its molecular, cellular and global aspects. However, I believe that the findings are in line with my overall proposal that there has been a fundamental shift in human brain structure and function that underpins the shift in human consciousness that occurred as we evolved from apes and is repeated as each human individual develops, and I will continue to discuss evidence for this later in this book.

In summary, interesting new evidence is emerging, both about how brains function at the molecular and cellular levels, but also as global entities, and also about what makes human brains different from those of other species. Yet I believe we are still only scratching the surface of what is going on in both these respects. This might seem a real problem for the model of human consciousness that I am seeking to develop in this book, but there are other ways to explore this issue than through neuroscience. In the next chapter, I want to look at how we can develop important new insights about the nature of human consciousness by focusing once again on the role of tool use and language in the development of human consciousness, in particular how this fuelled human rationality.

THOUGHT AND REASON 5

We saw that Aristotle viewed the ability to reason as central to specifically human consciousness. However, what is reason and what evidence is there that it is such a specifically human characteristic? To answer this question, let us return to the role that tools and language have played in human evolution and then think through carefully how this has affected the way we view the world around us, including the other human beings within it.

Turning to language first, one thing to reiterate about human language is that it is far more than just a means of communication. In particular, in contrast to the sounds that other species use to communicate with each other, human language is symbolic. What this means is that the words we use have no necessary connection to the objects they signify.

Of course, this is not always strictly true. For instance, the words that humans use to express the sounds that particular animals make do generally resemble the actual animal sounds, although it is interesting even here how much such sounds vary in different languages. So when I met my Portuguese wife-to-be, I found out that while children in Britain learn that dogs make the sound 'woof woof' and cockerels 'cock a doodle doo', in Portugal it is 'ão ão' and 'cocoroco'. The fact that pigs go 'oink' in English, 'boo boo' in Japanese and 'nöff-nöff' in Swedish further demonstrates the international variation, just for one type of animal.[1]

Yet, despite exceptions, the relationship between a word and the object it signifies is generally arbitrary, which has the consequence that the word for any particular object can be completely different between different languages.[2] Think, for instance, of the word for table in English, compared to that in Spanish, mesa; these words bear no resemblance to each other at all. In fact, in this case, both words derive from Latin, the original words being tabula and mensa, which originally referred to quite different types of objects, illustrating another important point about human language, which is that it evolves. Importantly, this combination of an arbitrary relationship between words and the objects that they signify, and the fact that words evolve, means that there is almost no limit to the words available to human beings as they search for ways to express themselves to others.

The power of human language is only partially related, though, to the complexity of its vocabulary. Equally important is that language follows a specific grammatical pattern. The mention of grammar often triggers in us negative memories of school and being asked to define the difference between nouns, adjectives, verbs and adverbs, or being expected to learn by rote the different verb endings in a foreign language. However, grammar means far more than the specific rules of a particular language. Instead, it is central to the way that human language allows us to specify the time, place and other features of a situation.[3]

Language instinct

Just how specific a human characteristic grammar is was shown by studies in the 1970s that attempted to teach human language to great apes. Because apes do not have the vocal apparatus to make the sounds associated with human speech, such studies sought to teach chimpanzees and gorillas sign language. However, while the initial results seemed promising, with some apes showing an ability to learn hundreds of signs and associate them with specific objects, animals, emotions and so on, in the outside world, none of the apes was able to combine symbols to generate complex, grammatically structured sentences.[4]

Such differences have led some commentators, such as the linguist Noam Chomsky, to suggest that this means that humans have a biological 'language instinct' for grammar. Further evidence that we have an innate ability to form grammatically correct sentences has come from the study of creole languages.[5] These languages, which get their name from the Latin word 'creare', meaning to produce or create, develop from 'pidgin' languages that arise when two groups of people who do not have a common language have to find a means to communicate. The first recorded example of a pidgin language occurred in China and got its name from the Chinese pronunciation of the English word 'business'. Pidgin languages are relatively simple and do not have a proper grammatical structure. However, if a pidgin tongue persists in a geographical region and children start to learn to speak it as their primary language, it develops into a creole that is grammatically correct.

The linguist Derek Bickerton saw this as evidence that humans are hard-wired to spontaneously 'know' basic rules of grammar. An alternative possibility is that children growing up in a region where pidgin is spoken 'borrow' grammatical rules from established languages, say English and the native language of that region, to create a creole.[6] In fact, as we will see later, environmental input also seems to play an important role in the development of language production and understanding in any particular developing human being, despite our apparent unique biological capacity for complex, grammatical languages.

The combination of an almost unlimited vocabulary and an instinct for correct grammar is what provides human beings with the ability to communicate with other humans in a precise fashion and with a creativity that, while often only used in everyday, mundane circumstances, can allow some humans to reach dizzy heights of novel artistic or scientific expression. Crucially, whatever the circumstances, language allows us to communicate in a rational way, for it allows us to express ourselves conceptually. Of course, many people expressing themselves to others may not be thinking, either explicitly or even implicitly, that this is what they are doing, but it's true of chitchat, as of learned discussion.

Inner thoughts

While most people reading this book may well agree with me about the unique characteristics of human language, my claim that language is also central to human thought might be more controversial. I will therefore try to explain further why I believe this is the case.

As I have already mentioned, a crucial aspect of the human thought process that distinguishes it from that of animals is its reliance on inner speech. In fact, a more accurate term would be 'inner symbols', because although words are one of the main symbolic forms that humans use to communicate and express ideas about the world, they are not the only one. For instance, deaf people tend to use sign language; indeed, studies of the thought process in deaf individuals have revealed that they appear to think in visual signs, not words.[7] More generally, spoken words are only one of the 'cultural' tools now available to humanity, others being music, visual art, literature and film. Therefore, to some extent, all of these symbolic forms might be part of the thought process.

Perhaps one of the biggest problems many people have with accepting the centrality of language to the human thought process is the difficulty in imagining life without language. Because our thoughts are so infused with the conceptual framework that language provides, it is almost impossible for us to imagine how it might feel to not have such a framework. Therefore, as a species, when we distinguish different colours or recognise that something happened in the past, as opposed to something that might happen in the future, or that we are an individual person and therefore distinct from other people, it can be easy to assume that other species have similar feelings. This may be one reason why children are very receptive to the idea that animals have human-like feelings, whether they are a character in an animated film or a pet cat or dog.[8] Even many adults may also not really think through the likely differences between human thought and that of animals; indeed, we should not underestimate the power of such assumptions, even in scholarly discourse.

A second reason why it may be hard for some people to view language, and symbolic forms more generally, as central to human thought, is the difficulty in assessing inner speech. I have already mentioned William James' comment that studying our individual thought processes is like 'trying to turn up the gas quickly enough to see how the darkness looks'. So there are definitely limitations to how far simple introspection can take us in terms of analysing our thoughts. Yet as I have also argued, there are indirect ways to examine the character of inner speech, ranging from studies of 'egocentric' speech in children, to asking people to record their thoughts, to what we can learn from studying dreams, art and literature. I will continue to mention such approaches throughout the rest of this book.

One final reason why someone might not agree with the centrality of inner speech for thought is that they may see it as simply incorrect. According to such a viewpoint, human beings do convey complex ideas to other humans through language, but this complexity is merely a reflection of what is actually going on deep down in their inner thoughts. Thus, far from it being language that structures human thought, it is instead our highly structured thoughts that find expression in outer speech and thought itself is quite distinct from language. While an interesting point of view, a problem that I have with this assertion is that it does not explain how thought processes that do not involve language could generate the conceptual framework that language does. In particular, such a viewpoint does not provide a material explanation for the uniqueness of human consciousness, because it fails to identify a material basis for the conceptual framework that underpins such a form of consciousness. Thus, by seeing thought as separate from language, it is ultimately an idealist vision of consciousness, something that I am trying to avoid in this book.

Although the view of consciousness outlined here sees inner speech, or more generally inner symbols, as central to human thought, I should stress that there is a lot more to our thoughts than these symbols. Lest this seems to contradict what I have said about the centrality of such symbols, I need to explain more what I mean in this respect. Essentially, what I have been arguing so far is that inner speech and other inner symbols act like a transmission belt between thought and outward expression. However, that does not mean that human thought and speech are equivalent. The reason for this is that ultimately all thoughts arise from a biological brain that, as we have seen, has much in common with other species' brains. While another central theme of this book is that the human brain has been transformed during evolution and continues to be transformed as each human being develops from babyhood, we cannot ignore these basic bio-logical features of our brains.

In particular, we cannot ignore the fact that our thoughts, at least at the deepest level of their biological origins, have important aspects in common with the thoughts of other species. Of course, that raises the question of what we mean by such thoughts. After all, I said earlier on that we could never know what it would feel like to be a bat, or any other kind of animal, because of the lack of any conceptual underpinning to their consciousness. However, given that we can feel pain, pleasure, fear or desire, we do have some insight into how these emotions might feel in an animal. In fact, and I will go on to explore this later, even the most basic human emotions have been transformed within our symbolic minds. As they are still based on similar brain structures and neurotrans-mitters as those in other species, any account of the human thought process must take this into careful consideration. In addition, we will look at the possibility that a sophisticated ability to categorise may exist in some non-human primate species, despite their lack of a language capacity.

Technological influence

While language is key to what gives human consciousness its rational character, we should not forget the role that our capacity for tool design and use plays in the formation of a rational mind, both in terms of evolution and individual human development. Here, though, we face a challenge, because while there is a clear link between language and reason, in that we use words when we put forward rational arguments, and, as I have argued, the grammatical struc-ture of human language is key to being able to make such arguments, the link between reason and the tools we use to transform the world is more indirect.

Yet there is ample reason to believe that systematic design and use of tools by humans predates our ability to communicate through complex language, so it may be seen as more fundamental in what first began to distinguish us from apes. In fact, archaeological studies have shown that our proto-human ancestors seem to have been designing and using sophisticated stone tools even earlier than had initially been realised, with the most recent studies having uncovered tools as old as 3 million years.[9] So how might this have affected our powers of reason? One thing to say on this score is that tools have become mediators between ourselves and the outside world for so long in our species' evolution, and also in the proto-human species that preceded *Homo sapiens*, that they have surely left a mark on the way we humans express ourselves about the shape of that world. Perhaps one of the most interesting aspects of human tools is the way that they evolved from generation to generation, a process that has dramatically increased in modern times, and changed us in the process.[10] While the development of human language has undoubtedly affected human technology, by increasing the sophistication of human imagination and creativity, we should not underestimate the degree to which new technologies can transform language.

We have only to think of how revolutionary the invention of reading and writing were in terms of increasing the human ability to communicate, as well as their effect on the development of civilisation. Recently we have seen how the internet has transformed our means of communication. Importantly, these changes affect the way that humans reason. Such changes have occurred within recorded history, so it is easier to assess their impact on human consciousness, but it seems perfectly likely that other major changes in tool use in prehistory also had important effects on language and the human thought process.

So both language and technology have had a big impact on human rationality. Although we can outline these effects at the outward level, what remains much less clear is how this connects with what is going on inside our heads. Therefore, with that in mind, and following my general goal of developing a material explanation of human consciousness, it is time to connect what I have been saying with specific aspects of consciousness. To begin to do that, I now want to look at perception: how we perceive the world, how similar perception in other species might be to this and what may be different in humans.

THE SENSUAL WORLD 6

A central feature of life on Earth is that all species, from unicellular bacteria to complex multicellular organisms like ourselves, have evolved ways to detect changes in the environment and respond accordingly, such activity being crucial to their survival. Demonstrating the unity of life on our planet, it is remarkable how similar, at least at the level of molecular biology, mechanisms of sensing the environment are between species. Such unity can be demonstrated by comparing the senses of a bacterium and a human.

As I mentioned previously, bacteria sense the environment through proteins on their surface called receptors. One class of receptors in bacteria that plays an important role in sensing changes in the environment is the rhodopsins. These respond to light, but also to other sensory stimuli. Remarkably, studies have shown that a rhodopsin protein also plays a central role in sensing light in the human retina,[1] and other proteins called G-proteins that evolved from rhodopsins[2] mediate our senses of taste and smell. If we needed a reminder of the essential unity of life, the demonstration that similar proteins fulfil the same functions in sensory processes in both bacteria and humans surely provides it.

Despite the similarities in the receptor proteins that both we and bacteria use for sensory purposes, a major difference is that such receptors in humans are in specialised cells that form only the first component in a chain that stretches all the way to the brain. In the eye, the sensory cells are called rods and cones; the rods are the most sensitive, but only the cones can sense colour. The signals generated in these cells travel along the optic nerve to the thalamus and then to the visual cortex, where they are used to generate the images that make up our visual impressions of the world.[3] Explained in this way, it might seem that our visual system is much like a digital camera that takes images which are then conveyed to our computer-like brain, where we view them. The actual situation is far more complex.

Visual input

To some extent vision is a bottom-up process. A major step forward in our understanding of this aspect of vision was made in the 1950s by the neuroscientists David Hubel and Torsten Wiesel.[4] They studied how the brain is involved in the act of seeing by recording the activity of neurons in the visual cortex of cats that were exposed to different stimuli on a screen in front of them. Their studies showed that individual neurons of the visual cortex respond not to complex shapes as such, but to simple lines and edges. As some neurons respond to vertical lines or edges, others to horizontal ones and others to ones at angles in between, the combination of these different neurons allows us to see complex shapes. In fact, this is only one aspect of how the visual cortex works. Therefore, while one part of this brain region receives signals relating to an object's shape, another area receives ones about its colour and another one about whether the object is moving.

Note, however, that none of this is a given fact. Other investigations by Hubel and Wiesel showed that covering up a kitten's eye at birth led to an inability to see with this eye, even if it was uncovered after three months.[5] Later studies by neuroscientists Colin Blakemore and Grahame Cooper showed that kittens kept in enclosures painted with only vertical lines never learned to see horizontal ones, for instance, a chair seat, while kittens only exposed to horizontal lines could not make out vertical ones, and would therefore bump into a chair leg.

Not surprisingly, studies like these have been controversial, but they have revealed many important aspects of the visual process that are directly applicable to human beings. For example, they explain why people who lose their sight soon after birth by some accident, like damage to the cornea of the eye, but have the defect corrected later in life by surgery, do not suddenly regain their sight, but have permanent distortions in their ability to see.[6]

Perhaps the most surprising aspect of what we are learning about the way the senses work is how much is determined by our brains in an apparently top-down fashion. I mentioned previously that visual signals go from the eyes to the thalamus, which acts like a relay station, and from there to the visual cortex. Such a flow fits with the idea of the cortex as a receiver of signals from the senses. However, what is interesting is that studies have shown that ten times as many signals flow from the visual cortex to the thalamus, at least as assessed by the number of neural connections in the forward and reverse directions.[7] The visual cortex itself also communicates with 'higher' centres of the brain like the prefrontal cortex and brain regions linked to memory.

Instinctual response

One type of input from the brain that is highly important in many animals is what we tend to categorise as 'instinct'.

Last year, my family and I acquired a kitten, whom we named Maya, and she soon showed a fascination with string and similar shaped objects, to the extent that we had to stop her playing with electric cords. So what might explain this fascination? Possible explanations are that the string resembles a mouse's tail, so it represents a likely piece of prey, or maybe in the wild cats might have to attack and kill potentially harmful snakes, or perhaps it is just the case that moving long, thin objects are more likely to catch a cat's eye. However, I think most people would accept that Maya's actions are probably part of some instinctual, stereotypical response and this raises the interesting question of whether human infants might similarly have such instinctual responses.

One strong reaction that a visual stimulus might be expected to produce in an infant is fear. Indeed human infants from about eight to ten months of age generally begin to notice the presence of snakes and spiders, avoid heights at the edge of a drop-off and withdraw at a stranger's approach. This has led to claims that such responses are based on hard-wired instinctual responses that have evolved to protect human infants from animals, places or other humans that might pose a threat to the infant's survival.

Yet a recent study by psychologists Vanessa LoBue and Karen Adolph found no evidence that human infants' responses to spiders, snakes, drops and strangers are either universal or primarily fearful.[8] Instead, the responses were shown to be highly variable, driven initially primarily by curiosity rather than fear, and were very dependent on the context in which infants encounter such stimuli, for instance, fearful reactions by parents or carers. LoBue and Adolph suggest that far from being maladaptive, the extraordinary curiosity that human infants show in their encounters encourages them to explore new things, while maintaining the flexibility to develop a fear if they discover something truly threatening.

The unusual flexibility that human infants show in their learning responses is mirrored by what is going on in the developing human brain. Studies have shown that both at the molecular and cellular level and at the level of gross structure, the human brain is far more 'plastic' – that it is affected in its development by the environment – than brains of other species, even those of our closest biological cousins, great apes.[47] As a result, the developing human brain is far more primed to respond to new experiences than the brains of other species, which undoubtedly affects our perception of the world around us.

Such a receptivity has a very important social consequence given what I have already said about the role of language, as well as other rapidly evolving 'cultural tools', like music, art and new technologies, in the development of human consciousness. This means that human infants are not only highly receptive to learning the language symbols that will eventually become the primary way that they communicate with other humans, and which are also critical to the development of human self-conscious awareness, but they are equally receptive to the many different technologies that characterise human society.

At this point, you may be wondering how all of this specifically relates to perception. To answer that, I would now like to look at evidence that human perception is uniquely different from perception in other species, in being a highly mediated form of perception.

Mediated perception

To some extent perception in all species is mediated. A bacterium does not respond directly to chemicals that signify food or danger; rather these are registered as changes in the concentration of ions within it. As we have seen, a cat's brain does not receive photographic images of a mouse or a dog on its visual cortex, but rather these register as a series of lines or edges in particular neurons, while neurons in other parts of the cortex respond to specific colours or movement, when a potential prey, or predator, comes into view.

While sensory experiences in humans can also be influenced by our evolved instinctual responses, and be affected by our learned experiences through individual encounters, a unique aspect of human perception is how it is mediated by tools and technologies. To some extent this can involve a literal mediation of perception by a tool. Think, for instance, of the way that photography affects the ways we view the world. Of course, our eyes can also see a scene that is captured by a camera's lens. Yet increasingly if we want to share such scenes with relations, friends and colleagues, we send digital images or videos to these individuals by e-mail, text or WhatsApp, or post them on Facebook, Instagram or TikTok. Challenging the notion of what we mean by 'reality', it is not at all uncommon for images to be modified prior to posting on social sites, for instance, altering the colours of a dish in a restaurant, the hue of a landscape or the intensity of a sunset.

However, are there also other ways in which our uniquely human characteristics have affected our perception that are more directly connected to brain function? A central way is linked to what I have said about the importance of language for human consciousness and how language can be viewed as a 'mental tool' that transforms human brain function. I have also mentioned how a unique feature of human language, in comparison to the communication systems of other species, is that only human language provides a means for the expression of abstract concepts such as time and place, self versus other and so on, and through the medium of inner speech this brings abstraction into our inner thoughts. However, such an ability to think in abstract terms also affects the way we perceive the world.

One example of how our ability to abstract affects our perception is the concept of colour. In fact, even the colours that make up the visible spectrum only represent a tiny fraction of the wavelengths of light that exist in the world, and it is testament to our modern technologies that we can indirectly visualise ultraviolet (UV), infrared or X-ray radiation.[9] Yet even the different colours themselves are as much an abstraction of the human mind as real entities.

Of course it is true that the different wavelengths of light exist outside the brain and also that our biological senses have evolved to detect such different wavelengths. Therefore the cone cells in our eyes come in three forms, which are often labelled red, blue and green cone cells, corresponding to what are known as the three 'primary' colours of light, but in reality represent particular wavelength ranges within the light spectrum. It is also true that with the help of such cone cells, many other species can distinguish different wavelengths of light. Not that we should assume that such other species' eyes see things the way we do. So while a cat's eyes also contain rods and cones, they have far more rods than we do, and so they see much better in the dark, but are worse at distinguishing the different regions of the visual spectrum.[10] Cats also distinguish blue and yellow light, but less so red and green.

With some subtle differences, both a cat's eyes and ours have evolved to discriminate between the different wavelengths of light. Yet only our species, I would argue, have a concept of colour, because only we can conceptualise reality. So how does this work?

At the heart of conceptual thinking is the ability to group things hierarchically. In terms of colour, this means humans have an overall concept of colour, with the individual colours subordinate to this. Similarly, we have an overall concept of shape and subordinate to this are specific shapes, for instance, circles, squares, triangles and so on.

Language is central to conceptual thinking because it allows us to describe concepts such as colour and shape, as well as to name individual colours and shapes, and also to employ even more general terms, such as 'characteristics' or 'properties', which link colours and shapes. However, conceptual thinking goes deeper than this, which is illustrated by another unique human capacity – the ability to produce and appreciate visual art. The ability to draw comes so naturally to our species that it is a great delight for a parent when a child first begins to create pictures, but it is also not surprising, being a stage nearly all children go through.[11] Although initially such attempts may be indecipherable squiggles, we rapidly notice when these metamorphosise into something recognisable, like a stick figure for a person.

Obviously drawing does not require words, although language can accompany it, but it does require an ability to conceptualise for the simple reason that it involves an ability to abstract. Such abstraction is shown by the fact that we can recognise a series of crudely drawn lines as a person as equally as we do a finely drawn, or photographic, image. This is important in terms of perception because it means we also perceive the world as an abstraction. That such a capacity is a uniquely human ability is shown by the failure of attempts to teach apes representational drawing.[12] This means that even such close biological cousins must perceive the world differently, because of their inability to abstract.

Visual art is possible for humans because our brains have not only been transformed by language but also by symbolic ways of representing the world. However, how does symbolism transform the brain? This is an issue I will continue to explore throughout this book, and as part of that exploration, it is now time to consider memory, for while our senses constitute a central way of interacting with the world, we also need to continually remember and learn from those interactions, both for survival and for the development of our consciousness.

LEARNING AND MEMORY 7

Learning and memory are as central to survival as being able to sense and respond to changes in the environment. Evolution by natural selection may be thought of as a form of species' learning, given that, within a species, variant individuals with particular biological characteristics can be best suited to survival in a particular environment, and therefore are more likely to pass down the genetic basis of their characteristics to their offspring.

Eventually, the characteristic that was only initially found in one individual will become representative of the whole species. A good example of this in unicellular organisms is the way that bacteria can become resistant to antibiotics during treatment for an infection. It only takes one bacterium in a million to gain a mutation that allows it to survive antibiotic treatment, and then to divide repeatedly until it represents the whole of the bacterial population, which is now antibiotic resistant.

A similar type of differential survival over generations led to the evolution of a long neck in giraffes. Originally, the giraffe's ancestors had short necks, but one variant's neck was slightly longer and this allowed it to reach the leaves on the tallest trees and become more likely to pass on its genetic differences to its offspring. Such examples can be seen as equivalent to the species learning how to survive in a particular environment. In an extension of this, human beings have evolved a unique form of social learning that is due to us being able to pass down spoken and written information to future generations. In general, however, at least when people talk about biological learning and memory, they tend to mean something that happens in an individual. So how widespread is this process in nature?

Synaptic changes

In multicellular species with nervous systems, a common theme is that, at the cellular level, learning and memory involves lasting changes in the connections between neurons. I mentioned earlier that neurons are connected together by cellular gaps called synapses, the neuron before the synapse being known as the presynaptic neuron and the one after it being the postsynaptic neuron. I also mentioned that when an electrical impulse known as the 'action potential' arrives at the end of a presynaptic neuron's axon, it triggers the release of a neurotransmitter from this axon, and the neurotransmitter diffuses across the synaptic gap and attaches itself to receptors on the dendrite of the postsynaptic neuron. Depending on the type of neurotransmitter and its receptors, this can either activate an action potential in the postsynaptic neuron or inhibit such an impulse; neurons are thus referred to as activating or inhibitory, with 10 to 20 per cent of neurons being inhibitory.

At the cellular level, memory involves structural changes in the postsynaptic neuron that make it more likely to respond if it is stimulated repeatedly by the same presynaptic neuron.[1] Such changes involve an increase in the number of receptors on the dendrite of the postsynaptic neuron, but also long-lasting changes in the neuronal 'cytoskeleton' – a complex, dynamic network of interlinking protein filaments that maintains the cell's shape and regulates the movement of substances within it – and even changes to the genes the neuron expresses, thereby helping reprogramme it as part of the learning response.[2] While initially learning was thought to involve only neurons, recent studies have shown that the brain's other type of cells, the glial cells that I mentioned earlier, play key roles in the alterations in postsynaptic neurons that underpin memory.[3]

Findings like these help to explain how memories can be stored at the cellular level: neural circuits responding to novel information from the environment store this information as material changes in the structure and activity of the neurons in that circuit. However, although we have a good understanding of how memory works at the cellular level, much less clear is how memories are stored in the brain as a whole, both in terms of which regions play a central role in this process and how they interact within the brain. Slowly though, through a variety of approaches, we are beginning to gain such understanding.

Brain lesions

Sometimes the role of particular brain regions in a mental process has become clearer, not from animal studies but from brain injuries in human individuals. We saw earlier how in the 19th century individuals with severe language defects were subsequently shown after autopsy to have lesions in particular brain regions, and this led to the discovery, for instance, of Broca's area, which seems to play an important role in language production.

In this particular case, it was syphilis that caused the lesion, and certain infectious diseases, both bacterial and viral, may have devastating effects on brain function. However, brain injuries can also occur when surgery goes wrong, as in the case of a US citizen called Henry Molaison.[4] As a teenager in the early 1950s, he began to suffer epileptic seizures that were so bad he had to quit school and could not hold down a steady job. It seemed that his seizures were originating from a brain region that contains a structure called the hippocampus. A surgeon, William Scoville, proposed to treat Molaison by surgically removing this brain region. At the time, little was known about the hippocampus' function, and the seizures had become so bad that Molaison agreed for the operation to go ahead.

The surgery did indeed result in a significant improvement in Molaison's epilepsy. However, it also had an unexpected, and terrible, side effect; from that point on, it became clear that Molaison had lost his ability to store short-term memories. Unable to recall daily events or recognise hospital staff taking care of him, and able to converse, but forgetting the content of a conversation only minutes after it had occurred, it was as if Molaison now existed in a perpetual present. Yet he could remember things that had happened to him in his youth, although only generally; thus he was unable to remember details about such events. Intriguingly, Molaison also showed an ability to learn new 'motor' skills. For instance, he was able to learn how to play tennis or trace the outline of a five-pointed star in a task that was made more difficult by the fact that he was only able to see the reflection of the star and his hand holding a pencil in a mirror. Later, Molaison had no recollection of learning such skills, as if only his unconscious mind was remembering them.

These aspects of Molaison's condition identified the hippocampus as an important region of the brain involved in memory; they also indicated which aspects of this process it might be particularly important for. Today we tend to distinguish the different types of memory as follows. First is episodic memory, which allows us to record specific events. Second is the type of memory that allows us to remember general aspects of our past, such as the school we went to or our parents' occupations. It also allows us to remember more significant events in the past, such as the Second World War. Finally, procedural memory allows us to learn new motor skills, like riding a bike or playing the piano.

The fact that Molaison could not remember things that had happened to him minutes before, but he could recall general events and also learn new motor tasks, led to the idea that the hippocampus plays an important role in episodic memory, but not in other forms of memory. Essentially, the hippocampus was viewed as being like a relay station that recorded memories transiently, that were then stored in a more permanent, and general, form, in the frontal areas of the cortex, while procedural memory was dealt with by the cerebellum. However, a number of more recent findings have challenged this viewpoint.

Recent studies of the role of the different brain regions in memory performed in mice have indicated that both memory formation and the storage of memories are based on a more complex, and interactive, relationship between the hippocampus and the 'higher' regions of the cortex than had previously been imagined. Such studies have used a revolutionary new technique called optogenetics.[5] I mentioned earlier that some unicellular organisms can sense light via receptors on their surface membranes that allow a flow of ions into the cell in response to the light. This then activates proteins within the cell that stimulate a response, such as a movement towards the light. Optogenetics uses genetic engineering to express such light-sensitive receptors in neurons in the brains of mice.

Because the electrical signals that flow along neurons are powered by ion flows, by inserting microscopic fibre-optic cables into specific regions of the genetically engineered mouse brain, it has been possible to stimulate such neurons in a living mammal by shining light on to different regions of the genetically engineered mouse brain. As different types of light-sensitive receptors can either activate or inhibit neurons, this allows scientists to either activate a particular region of the mouse brain or suppress its activity.

Optogenetics has been used to study a variety of processes in the brain, but one particularly fertile application has been the study of memory. Scientists have shown that it is possible to artificially stimulate memories of something that has happened to a mouse in the past, for instance, a painful stimulus, and also repress memories. Some studies have focused on exploring which regions of the brain are involved in the memory process.

One such study has shown that memories are formed simultaneously in the hippocampus and the prefrontal cortex, the latter region being one that is thought to play key roles in 'higher' mental activities in humans, such as planning, reasoning and so on. However, memories in the prefrontal cortex remain 'silent' for several weeks, as demonstrated by the fact that neurons in this region show clear signs of remembrance of a painful stimulus when activated artificially by optogenetics, but no such activation during natural memory recall.[6]

Such findings suggest that memories are formed in both the hippocampus and the prefrontal cortex, but initially the prefrontal memories are silent and only later become cemented as long-term memories, while the hippocampal ones are erased. However, the nature of this long-term cementing action remains to be determined, and the example of Henry Molaison suggests that, at least in human beings, the hippocampus cannot be simply bypassed during the normal formation of memories. In fact, another study has suggested that the hippocampus acts as a 'convergence zone' that pulls together different bits of information stored in the cortex that relate to different aspects of an event into a coherent whole.[7]

Memory concept

If such studies have revealed some potential mechanisms that underlie the formation and storage of memories in an animal brain, how does this relate to human memory? Here it is worth mentioning a study performed in human volunteers. Led by neuroscientist Rodrigo Quian Quiroga, this involved individuals with epilepsy and, in this case, after receiving a local anaesthetic, their brains were exposed and microelectrodes implanted into specific neurons. The idea was to try and determine where in the brain the epileptic seizures were being triggered so that these areas could be surgically removed without damaging some important cognitive function, as happened to Henry Molaison. While undergoing the surgery, the individuals also took part in an experiment that involved them being shown different photographic images of animals, objects, landmark buildings and celebrities, while recording microelectrodes were implanted into different neurons within their hippocampus.

A surprising outcome of this study was that particular neurons only became activated in response to a specific image. So one neuron was only activated in a volunteer when they were shown an image of Bill Clinton, the former US President. In another case, a neuron only became activated in response to an image of actor Jennifer Aniston.[8] Other neurons in the hippocampus responded to images of famous landmarks like the Eiffel Tower. Illustrating a dynamic aspect to memory, when a volunteer whose hippocampal neuron had responded to a picture of Jennifer Aniston was shown a picture of her alongside the Eiffel Tower, the neuron soon began to respond to pictures of this landmark on its own.

What is going on in the hippocampus to explain these findings? One possibility is that the hippocampus somehow acts like a kind of memory-laden matrix with all the objects we have remembered in the past mapped onto it via different individual neurons. Although that may seem implausible, it is worth mentioning other findings from studies by the neuroscientist John O'Keefe performed in the early 1970s, which showed that the hippocampus can be viewed as a kind of map, at least for spatial memory. The studies, which involved implanting microelectrodes into the brains of rats and monitoring their activity while the rats moved around a box were initially meant to try and identify regions of the brain involved in regulating movement. However, when he inadvertently implanted a microelectrode into a hippocampal neuron, O'Keefe was surprised to see it activate when the rat moved into a particular part of the box. Further work showed that the hippocampus contains neurons now known as 'place-cells' that help create a memory of the environment so that, as O'Keefe explained it, 'if you put them all together, you could have something like a map'.[9] These days we might describe them as akin to a GPS tracking system.

Subsequent studies have shown that place-cells also exist in the human hippocampus and that the spatial component of episodic memory in human beings may be encoded by these cells. However, when it comes to other aspects of memory, we need to find other explanations than the idea that the hippocampus is merely a kind of map of objects and people that we encounter. One reason is that there are more abstract aspects to memory than simply image recognition. To some extent this was demonstrated by the study mentioned above, which showed images to volunteers and measured the response of particular hippocampal neurons.

In one volunteer, a neuron was activated by an image of the actor Halle Berry, and also by one of her playing the masked character Catwoman and even by a picture on which was printed the words 'Halle Berry'. This suggests that far from this particular neuron's response being merely to a photographic image, it seems rather to be 'responding to the concept, the abstract entity, of Halle Berry', in the words of Quiroga, who led this study.[10] I have already argued that the ability to think in terms of abstract concepts is both uniquely human and centrally linked to our human capacity for language. Therefore, somehow we must link this aspect of human memory to language.

In addition, I believe it would be a mistake to only focus on the hippocampus' role, for memory also involves other parts of the brain such as the prefrontal cortex, as well as the cerebellum. In what follows we will therefore need to tie all these different aspects together. I will be aiming to do that in subsequent chapters of this book, but for now I want to return to the fact that, while they have been transformed by our species' capacity for language and technology, our brains are still biological entities that have evolved over millions of years. In particular, I want to consider what role emotions play in human consciousness and how their role is both similar, and different, to the situation in other animal species on Earth.

MIND CHEMISTRY 8

A central theme of this book is that the human brain has been transformed, both during our evolution from apes and as each one of us develop from babies within human society by our capacity for language. Expressed in this way, it might sound as if human consciousness is being viewed as a kind of word processor. Another viewpoint sees the human brain as essentially a circuit diagram. This view underlies the quest to identify the basis of human consciousness by mapping these neural circuits in the Human Connectome Project.

In fact, the human brain is neither just a word processor nor an electronic circuit, because it is also a biological entity that is the product of millions of years of evolution. This means that we must also consider its complex biology if we want to understand how such an entity can form the basis for each individual human consciousness. To some extent we have already looked at the structure and biology of the neurons and glial cells that together constitute the brain. I have also discussed the gross structure of the brain. Now, in this chapter, I want to say more about the chemistry of the brain and to consider how this relates to what are known as the 'emotions', for I would argue that our emotions are as much a part of our consciousness as is our ability to think, conceptualise and rationalise.

The word 'emotion' is derived from the French word *émouvoir*, meaning to move or stir up.[1] This fits with the strong physical disturbance that can accompany an emotional response. The fact that we can talk of lovers being drawn together by body chemistry shows a recognition of the fact that love, like other strong emotions, involves chemical changes in the body. But how does this relate to the human brain and to consciousness?

We have already touched upon one aspect of chemical action that is central to brain function. To recap, although signals travel along a neuron in the form of an electrical current, the transmission of signals from one neuron to the next involves the release of a chemical called a neurotransmitter at the end of the presynaptic neuron's axon. The neurotransmitter diffuses across the synaptic gap and stimulates a response in the postsynaptic neuron by attaching itself to a 'receptor' protein on this neuron's dendrite.

Brain messengers

This then is the basic mechanism whereby signals travel through neural circuits. However, the situation is complicated by the fact that there are many different types of neurotransmitters, each with a different effect on the neurons that they stimulate.[2] Some of these chemicals are relatively well-known to the public, such as dopamine and serotonin, others like glutamate and gamma aminobutyric acid (GABA) less so. However, all play important roles in the brain.

One of the reasons why dopamine and serotonin have become household names is that they are commonly associated with disorders of the brain and mind. A lack of dopamine in one brain region is therefore thought to be a primary cause of Parkinson's disease, while too much of this neurotransmitter is proposed to be responsible for the mental disorder schizophrenia. A different mental disorder, clinical depression, has long been thought to be due to a deficit in the neurotransmitter, serotonin. In fact, there is increasing criticism of the idea that mental disorders are primarily due to too much, or too little, of a particular neurotransmitter in the brain. Instead, focus has shifted to looking at how abnormalities in the structures of neural circuits or in the dynamics of how different parts of the brain interact may underlie mental disorders, and while these may involve changes in the levels in the brain of certain neurotransmitters, this may be only part of the cause of the disorder.

While different neurotransmitters are known to play distinctive roles in the brain, what still remains far from understood are their precise roles in different brain regions. What is clear is that a particular neurotransmitter can have many different roles. Dopamine is involved in the control of movement, which is why degeneration of a brain region called the substantia nigra, which has an abundance of neurons containing this chemical, occurs in people with Parkinson's disease, causing first tremors and then stiffness or slowing of movement. Dopamine also plays an important role in reward and pleasure-seeking behaviour, a feature of its action first identified by the neuroscientist Wolfram Schultz. He was studying the behaviour of neurons in the monkey brain and noticed that certain neurons released dopamine when the monkey was given a reward. He also showed that this release pattern changes as the animal learns how to respond to receive the reward and that learned cues could trigger changes even without a reward. Of particular interest was the fact that the pattern of release could distinguish between a reward that is being received and a reward that is predicted to be received in the future.

This shows that quite sophisticated aspects of behaviour can have a chemical basis. Subsequent studies have mapped the regions of the brain in which dopamine-releasing neurons are located. In general, these seem to be very similar in a monkey or human brain. However, a recent study has shown that a gene called TH that is involved in the production of dopamine is active in humans – but not other primates – in the neocortex region of the brain that is particularly associated with 'higher' forms of thinking, that is ones associated with reasoning, planning, abstraction and so on.[51] This suggests that during the transformation of the human brain that has occurred during our evolution from apes, unique roles may have emerged for particular neurotransmitters in human consciousness.

I mentioned earlier that a major change in the way we view the brain involves an increasing recognition of the role played by glial cells. Formerly, glial cells were thought to have a passive, supporting role, their very name reflecting the opinion that they behaved as the 'glue' that held the neurons in place. This negative bias extends to the general name that is given to brain chemicals – *neuro*transmitters. This reflects the focus in neuroscience on the neural circuit, linked together by synapses, and therefore the chemicals released by neurons into that synaptic gap. Yet just as we are now learning that different types of glial cells play a variety of important, and active, roles in the brain, so there is increasing recognition that while glial cells are not joined together in electrical circuits, they do release chemicals, and they also possess receptors on their surface that respond to chemicals. This therefore needs to be factored into any discussion about brain 'chemistry'.

If all of this hopefully conveys a picture of the emerging complexity of brain chemistry, how does this relate to our emotions and their role in shaping human consciousness? Here we face a curious contradiction in the way that emotions are viewed in human society. For although we tend to think of emotional responses like anger, fear or sexual desire as demonstrating the 'animal' side of our natures, we also elevate emotions like 'true love' to the pinnacle of human responses and something not possible in an animal, while seeing a negative one like envy as a baser emotion in a moral sense, but similarly specific to humanity. So what is it about emotions that allows them to span such a range of feelings, signifying our basest animal nature but also the highest refinement of our species?

Animal origins

One feature of emotional responses is their instinctive nature and stereotypical form. As humans, we recognise a strong emotional response in ourselves or others by the fact that it can often seem to happen automatically or without much rational thought. A good reason for this is that if an organism is confronted by a predator or other potentially life-threatening situation, the immediacy of the situation means that it does not necessarily pay to spend too much time weighing up the pros and cons of one type of behaviour over another. Instead, what may be most useful in this situation is a rapid, hard-wired response. Similarly, an intense emotional attraction to a member of the opposite sex of the sort that in humans we often refer to as 'blind' love, can help ensure the propagation of the species.

An interesting question when considering human consciousness is to what extent this reflects our animal origins and to what extent it involves uniquely human features. It is generally agreed that animals can show signs of anger, fear and sexual desire. However, an emotion like envy is surely something that only humans can feel, as it is based on an ability to imagine possessing something that we ourselves do not have, but another person does.

We might have more debate about whether animals can feel true love or be jealous, as a pet cat or dog might be seen as loving its owner, or feeling jealous if that owner shows affection towards another pet. However, in ascribing such feelings to pet animals, there is the danger that we are showing anthropomorphism, the attribution of human traits, emotions or intentions to non-human entities. In fact pets are an interesting category of animal, as cats and dogs have specifically evolved to live with human beings. So some of what seems like human-like behaviour on their part – such as a dog looking guilty when caught doing something wrong or gazing wistfully at their master or mistress if they desire something tasty to eat from the dinner table – have evolved as hard-wired behaviours that have developed as part of making such species good companions for humans.[3]

While we can separate emotions into more animal-like ones and those that we see as unique to human beings, the central theme of this book is that human consciousness has been transformed by language and tool use. A consequence of this is that all emotions, even the most apparently 'base' ones, must also have been transformed in ways that make them fundamentally different from equivalent behaviours that we can observe in animals.

One has only to think about the difference between someone starving to death in the desert compared to a person having a meal in an expensive restaurant to realise that context is everything, even for as basic a feeling as hunger. In fact, practically any emotional response that we have is affected by context. Disgust, anger, fear, desire, love, happiness, jealousy, envy, sadness, surprise and boredom are emotions that can be found across human societies, but the events that trigger such feelings can be very different in such societies and I would argue that this affects the responses themselves.

Novel connections

What do we know about the regions of the brain involved in emotional responses? The amygdala is a brain region that is particularly associated with the emotions as it has been shown to be involved in fear and anxiety but also pleasure.[4] The amygdala is part of the limbic system that has ancient evolutionary origins and for this reason is sometimes known as the 'lizard brain'. An important feature of the human brain that is becoming increasingly obvious from recent studies is the presence of novel connections between different brain regions, so we should expect emotional responses to be influenced by these interconnections. In particular, the ability to think conceptually and rationally means that humans have a unique ability to override their emotions and we should expect this ability to be reflected in greater control over brain regions such as the amygdala by other regions more associated with the 'higher' forms of consciousness, such as the prefrontal cortex.

So far, our discussion of the emotions has focused on the brain. Yet there is increasing recognition of the role played by the rest of the body in emotional responses and in consciousness more generally. This is because of evidence of neural connections between the brain and different parts of the body and because recent studies have shown that the organs themselves can send out chemical signals that can influence other bodily regions and the brain. Such studies challenge the idea that information only flows from the brain to the body. Instead, it is becoming clear that our organs have an important role to play in sending messages to the brain. Showing the complexity of this interaction, many chemicals that have been identified as having important roles in the body also operate in the brain.

The hormone oxytocin plays important roles in pregnancy, giving birth and the production of milk. Yet oxytocin is also active in the brain, influencing behaviours including social recognition, bonding and parental behaviour.[5] It also plays an important role in orgasm and its concentration in the brain increases when a person falls in love. For this reason, oxytocin is sometimes referred to as the 'love hormone', but this term is misleading as the chemical can also intensify memories of bonding gone wrong, such as in men who have poor relationships with their mothers. Oxytocin can also make people less accepting of individuals viewed as outsiders. This is another example of how the context of a social interaction can influence the type of emotional responses associated with it.

The important role that emotions, which are based on chemical changes in the brain and body, play in human consciousness is one reason why it may be difficult to mimic such consciousness in a computer. I will be looking at differences between brains and computers later in this book. However, now I would like to consider a different issue at the heart of human consciousness, namely if consciousness really is just brain activity, how does this explain that intensely personal feeling each of us have of being an individual consciousness? To study this issue further, let us turn away from neuroscience and delve into some philosophy.

PHILOSOPHY OF MIND 9

In the last few chapters, I have highlighted some of the latest insights from neuroscience about the relative roles of neurons and glial cells, the interconnections between the different brain regions, the chemistry of the brain, the relationship between the brain and the rest of the body and how all of this might relate specifically to human consciousness. However, while we can identify all the ways that the biology of the brain might underpin consciousness, is this really the same as explaining what it means to be a conscious person?

One philosopher who is sceptical about such a claim is David Chalmers, for as I mentioned earlier, he has stated that no matter how much advances in neuroscience allow us to solve what he calls the 'easy' problems of consciousness, by which he means such things as understanding how we learn, store memories, perceive things, react to a painful stimulus or hear our name spoken in a crowded place, the 'hard' problem of consciousness, which means explaining scientifically that subjective feeling a person gets when they see a beautiful sunrise, savour the taste of a ripe mango, fall in love for the first time or muse over a brilliant novel, is a much harder, and possibly impossible, problem to overcome.

In fact, Chalmers is far from alone among philosophers in his scepticism about this matter. Yet, equally, there are philosophers who reject the idea that the hard problem is so insurmountable, or even that it exists at all. So is it possible to combine philosophical insights with those from neuroscience, link this to what I have said about the unique aspects of human consciousness and, in the process, move things forward both philosophically and scientifically? This is what I will be trying to do in this chapter, but first I want to consider more explicitly the different points of view within the philosophy of mind and how they relate to the matter at hand. In so doing, we first need to consider two very different ways of viewing the world, that is the opposed concepts of materialism versus idealism.

Materialism, also known as physicalism, states that everything in the world, from the vastness of the universe to our deepest inner thoughts, can be seen as different manifestations of matter and energy, without any requirement for supernatural forces. This viewpoint is the one most closely aligned with the natural sciences. Indeed, it is hard to imagine how any modern scientific subject would have anything but materialism at its core. Yet the study of consciousness seems an exception, in that some distinguished and influential philosophers of mind today espouse not materialism but its opposite, idealism. Actually, I should make it clear that some views I consider idealist are expressed by people who seem to believe they are putting forward materialist positions. To address this issue, in this book I will define an idealist viewpoint as an explanation that involves supernatural forces, but also supposedly materialist explanations that are not based on current laws of science.

Idealistic view

In its purest form, idealism was best expressed by the 18th-century philosopher George Berkeley.[1] A bishop of the Anglican Church of Ireland, Berkeley took a different approach to Descartes' dualistic take on consciousness, with its unhappy union of a mechanistic body and an immortal, supernatural soul. For unlike Locke and Hume, who sought to move beyond such dualism by proposing consciousness to be a purely material phenomenon based on life experience – what I have called the 'blank slate' view of the mind – Berkeley countered that only God and immortal souls exist, and because God is all powerful, he created the appearance of a material world existing outside our minds, but that this world is just an illusion. As a religious viewpoint, this may sound plausible, for if one believes in an all-powerful God, presumably anything is possible. Yet I find it surprising to find such idealism in some current philosophies of mind, apart from theology departments. Surely a central aspect of modern academia ought to be explanations that do not rely on the supernatural. Therefore, while looking at such explanations, and potential flaws within them, it is also worth asking why idealism holds such power within the philosophy of mind.

Let us start by looking at materialist explanations and why some philosophers have found them wanting. I have already mentioned Daniel Dennett and his view that any truly materialist account of consciousness has to reject the idea of anything equivalent to a homunculus, or 'little man', directing things from inside the brain or, for that matter, any sense of a specific place in the brain where 'it all comes together'. All this seems to me a correct view and yet I have also previously criticised the way that Dennett ends up falling back on the concept of 'memes' to explain how ideas are generated, which I see as an idealist viewpoint, as it views ideas as objects with a life of their own, unanchored to human beings.

Another philosopher who believes it is important to reject the view of consciousness as centred around some inner homunculus is Keith Frankish. Like Dennett, he believes consciousness is better viewed as a kind of illusion.[2] Note that Frankish is not denying that conscious experiences exist, rather that such experiences do not 'involve awareness of non-physical, private mental qualities, presented like a show to some kind of inner observer'. Instead, Frankish argues that consciousness is more akin to a 'news report rather than a theatrical show'. Importantly, 'this report isn't in a human language – it's in the brain's internal language of neural signalling – and it's not for the benefit of an "inner you"'.

Frankish does, however, still believe that consciousness allows introspection, the ability to look inwards and be aware of our own experiences. However, he argues that even this introspective capacity 'depends on sub-personal mechanisms of reporting and reacting ... Introspection is just another layer of information processing, this time directed onto other brain processes.' This means that even introspection is an illusion, because its 'impressionistic sub-personal reports on complex brain processes lead us to think that we have direct and infallible awareness of private mental qualities but we don't'.

Mental illusion

Such are some of the main avowedly materialist philosophical explanations of consciousness, but a number of influential philosophers are not convinced. For instance, Galen Strawson has called the idea that individual consciousness is an illusion, 'the silliest claim ever made'.[3] Strawson's main objection is linked to Descartes' great insight 'I think, therefore I am'. Whatever else we might doubt about the reality of the outside world, that feeling we have when we wake in the morning of immediately being aware of ourselves as a unified individual consciousness is surely a key certainty of life that is hard to square with simply being an illusion, as is the highly personal feeling of the many sensory experiences, thoughts, ideas, inspirations and setbacks that our minds undergo during our lives.

But if Strawson rejects the idea that consciousness is an illusion, what does he pose as an alternative? In considering this question, it is time to introduce the notion of 'panpsychism'.[4] This refers to the idea that consciousness is not just a feature of human beings, or more generally of other complex species, but a central characteristic of matter itself. I have already mentioned that the neuroscientist Christof Koch has proposed that consciousness is a feature of organisms stretching from human beings to simple bacteria. However, panpsychism goes even further, for if consciousness is a basic feature of matter, then this implies that not just living things, but also inanimate objects, have consciousness.

A confusing aspect in assessing philosophers of mind who espouse panpsychism is that many appear to believe their viewpoint is materialist. Yet such a belief is confounded by the lack of any scientific evidence for panpsychism. It rather seems the case that a perceived inability to explain human consciousness using current scientific methods has led to the claim that some novel and so-far undiscovered property of matter must exist that underlies consciousness. Stated like this, panpsychism can claim to be a type of materialism, just one we cannot as yet understand scientifically. But surely this means that it is an idealist view, at least as defined by my rule that materialism must follow the laws of current science.

In fact, there has been an attempt by the physicist Roger Penrose to provide a material explanation for consciousness based on the idea that this is a universal property of matter, but linking this to biological mechanisms.[5] Penrose, quite rightly in my mind, argues against the idea that human consciousness can be reduced to a computation. Instead, he has focused on the often-indeterminate nature of our thoughts. Seeking a source for this indeterminacy, he links it to another indeterminate natural phenomenon – the quantum world of subatomic physics. In particular, Penrose uses an insight from the field of quantum computing to suggest that isolated bits of information may remain in multiple states in the brain until coming together in an instantaneous calculation, called 'quantum coherence', making them act together in one quantum state. He believes this explains how, amid the buzz of unconscious activity in our minds, occasionally a thought, inspiration or feeling emerges from the general background noise and pushes into our conscious awareness.

If this seems rather like a leap in the dark, the flaws in Penrose's theory become particularly obvious when he tries to identify a biological substrate for his quantum effects. For, rather bizarrely, he picks upon a subcellular molecule, microtubules – long, thin proteins that form part of the cell's 'cytoskeleton' – as the bearer of such consciousness, in particular proposing that vibrations in such molecules could mediate quantum effects. Now there is no doubt that the cytoskeleton plays all sorts of important roles in the brain, for instance as a way of making more permanent the cellular changes that underlie learning and therefore providing one basis for long-term memory. However, to specifically focus on microtubules above other structures in the cell as the basis for consciousness makes little sense to a biologist like myself. A more fundamental problem I have with Penrose's viewpoint is that seeing consciousness as a basic feature of matter confuses a phenomenon that occurs at the highest level of organisation – the human brain – with those occurring at lower levels.

Emergent properties

An important aspect of Penrose's view of consciousness as a quantum phenomenon is how it treats the concept of 'emergence'.[6] This builds upon the fact that the natural world is organized hierarchically from the level of subatomic particles to the cell, organism, ecosystem, biosphere and so on. Each proposed hierarchical level has certain emergent principles that do not appear in the previous level of the organisation. As a counterpoint to the idea that complex systems can be only understood by reducing them to their basic units this is a useful point of view, but a problem with this approach is that it often neglects the importance of describing the distinct structural properties of different levels and thereby can lead to blurring of the mechanisms by which one level becomes another. For instance, in Penrose's desire to locate the root of consciousness in the quantum world, he ends up focusing on one arbitrary structure in the cell – microtubules – while ignoring the rest of neuronal and glial molecular and cellular structures, the organisation of the brain into different sub-regions and their interconnections via electrical and chemical communication, as well as the link between the brain and human social interaction, primarily through language – all key elements that I have identified as important for understanding human consciousness.

Betraying the panpsychic view of consciousness as an idealist vision is how easily it can slide into a mystical view of the universe. For if human consciousness is a direct product of the quantum world, it is not a great leap then to begin to see consciousness everywhere, even in the basic fabric of the universe, and indeed this is what some physicists now believe.

For instance, the physicist Gregory Matloff has recently argued that a 'proto-consciousness field' could extend through all of space and that stars may be thinking entities that deliberately control their paths.[7] Put more bluntly, the entire universe may be self-aware. In support of this idea, Matloff has noted that some stars appear to emit jets of gas in only one direction, an unbalanced process that could cause a star to alter its motion, and he has suggested that this may be a 'wilful' process.

It is most likely true that unexplained properties exist in the fabric of the universe that could explain the behaviour of universal bodies like stars, and perhaps also the discrepancy between the properties of the subatomic quantum world and those of the universe as a whole, something that current quantum and relativity theory fail to do. However, to characterise physical processes as 'conscious' blurs the boundary between mind and matter in a way that detracts from the task of explaining each in its own terms and is also potentially only a short step to belief in the supernatural idea that 'god is in everything'.

Whether or not panpsychic explanations seek to present themselves in a materialist, scientific fashion or end up being openly mystical, I consider them idealist explanations because they are not rooted in any kind of science that it is possible to test. Yet I have also mentioned criticisms of avowedly materialist explanations of consciousness that also fall short, such as the way Locke and Hume failed to explain how a unique and distinctive human personality is possible with their blank slate view of the mind or Daniel Dennett substituted memes, idealist entities, for a materialist explanation of the origins of human ideas. So how might we seek to provide a materialist account that fully recognises the complexity of consciousness and its individual, subjective qualities, but does not lapse into idealism?

My proposal is to build on what I see as the crucial distinction between specifically human self-conscious awareness and the more general type of consciousness found in other complex biological species and to link these two types to unconscious brain mechanisms. However, to do this, we need to return again to the importance of language and tool use for our species and ask how this has transformed the human psyche, not just in a general species sense but in an individual, and how this relates to brain mechanisms.

INDIVIDUAL AND SOCIETY

10

Human consciousness is based on what can at first seem like a contradiction. On the one hand, humans are a highly social species and I have already argued that we are also unique in that only our species is able to communicate conceptually through language. This has had major implications for human evolution, because concepts can also evolve, allowing us not only to pass on ideas to other human beings but also for those ideas to change through the generations. Then there is the other way that human society evolves, which is through the development of technology. Yet against this highly social aspect of our species, if there is one thing that seems to define any particular human consciousness, it is its highly individual character. It was this that led Descartes to state 'I think, therefore I am', and as we muse on our own inner thoughts, one of the most striking things about them is their highly private and personal character. So how do we square the social species with that apparent individuality?

In order to answer this question, we need to return to the link between language and thought, before we then try and connect thought with what might be going on within the brain. I have already argued that a central way in which human thought processes differ from those of animals is through the mediation of our thoughts by language. Inner speech, or some other form of inner symbols of the sort mentioned previously, is not just the way we transform thoughts into outer speech or symbols, but integral to thought itself.

This has implications for the structure of human thought, for just as human speech or sign language generally involves an interaction between two or more people, and the structure of human language reflects this, so inner speech or signs have social origins and a structure that reflects these origins. The philosopher Valentin Voloshinov recognised this when he stated that 'individual consciousness is not the architect of the ideological superstructure, but only a tenant lodging in the social edifice of ideological signs'.[1]

Social mind

In Voloshinov's view, human consciousness has as much a social as an individual structure, precisely because it is based around words and the grammatical rules binding these into an intelligible, conceptual framework, and which have evolved through previous interactions between human beings. Another crucial aspect of inner speech is that although we believe we think in monologues, inner speech's social structure, and the way our consciousness develops through social interactions as we grow into adults, means that in reality inner speech is a dialogue with different outside voices that have shaped our consciousness in the past. This dialogic aspect is I believe of the most underrated, but profound, aspects of specifically human consciousness and it is worth exploring its implications in greater detail.

For a start it undermines the view expressed by Descartes that 'I think, therefore I am'. Although the one thing we may think we can know about our consciousness is its unitary, individual character, this is partly an illusion. Yes, our consciousness is real, but there is nothing unitary about it, fashioned as it is from multiple voices. As I will explore in more detail later, in some types of mental disorder, even the illusion of a unitary consciousness can vanish as the personality becomes distorted or even totally fragments.

A second important implication of the dialogical structure of inner speech is that the patterns of external social conversation influence what is going on in our heads. Voloshinov proposed that although when we make an utterance, we may think that we are linking the words that come out of our mouths in an original and individual way, actually utterances are heavily dependent on 'speech genres'. People learn as they go through life that particular circumstances tend to produce specific types of verbal interactions. For instance, two strangers waiting at a bus stop may break the silence by talking about the weather, and to do so they will use phrases shaped by conventions of the genre 'talking about the weather'. Voloshinov thought that speech genres similarly operate in a variety of other situations.

There can be genres based on social events, such as the one just mentioned, but also ones that reflect different aspects of the speaker, for instance their gender, age or status in society. Speech genres are distinguished both by the kinds of words used and by more subtle facets of speech such as intonation. For instance, studies have shown that women's speech is traditionally more varied in its intonation, being both more questioning and exclamatory in its nature than men's speech, and is characterised by a rising intonation.[2] In contrast, men's speech tends to be more uniform and assured, with a downward intonation. Such differences are in line with the idea that women are more openly emotional than men, the latter being perceived as stiffer and more reticent. Of course, these differences may also reflect women's traditionally lower status in society, but notably such gendered differences in intonation seem to be becoming less obvious in societies in which women are beginning to play a more equal role at work and in society as a whole.

In general, speech genres impart a regularity to communication while remaining open to the shifting pressures of daily life, as well as being influenced by social and political change. Speech genres temporarily crystallise the relations between two speakers – their respective power and status, the purpose of the conversation, its subject and relations to other conversations. As such they can be highly ideological, for as Voloshinov put it, 'each word is ... a little arena for the clash and criss-crossing of differently orientated social accents. A word in the mouth of a particular individual is a product of the living interaction of social forces'.[3] Speech genres can also reflect changes in such forces. I will return in more detail later to the ways that language and social and political change are linked together, but for now, I just want to note that while speech genres mediate an individual's interactions within society, they also play an important role in human consciousness.

Black box

If human consciousness is at one level a dialogue between different voices, if we want to explain consciousness in material terms, we surely need to link the way that language structures human consciousness to the fact that such consciousness is occurring in a biological brain. Now it might seem obvious that any scientific explanation of human consciousness must be heavily rooted in an understanding of our brain's structure and function, but surprisingly this was not the case for mainstream psychology for a substantial part of the 20th century, due to the influence of a current within it known as 'behaviourism'.

Behaviourism developed from an observation made by the biologist Ivan Pavlov.[4] He was studying the process of salivation in dogs, which he measured using a tube inserted into the inner cheek that was connected to a recording device. Pavlov predicted that dogs would salivate only when food was placed in front of them, but he found to his surprise that they actually began to do so on hearing the footsteps of the assistant bringing them food. The dogs had learned to associate the footsteps with being fed. Pavlov also found that he could stimulate salivation by subjecting the dogs to any sound that they came to associate with the prospect of food. Moreover, if the food was removed, the dogs trained in this way still salivated in response to the sound. Pavlov called this learned response 'conditioned' learning, in contrast to the 'unconditioned' response that is the dog's instinctive salivation on smelling food. He speculated that conditioning must also reflect changes in the brain.

Pavlov's discovery represented an important step forward for our scientific understanding of the mechanisms underlying behaviour, not just of animals, but also of human beings. Different types of conditioned responses that can develop in people include taste aversions, fears, tensions and favourable feelings. A person may feel ill after eating a particular type of food and from that point on may develop an aversion to that food, even if the illness was coincidental and it was not the food that had made them feel poorly. More generally, people can become irrationally fearful of certain types of objects, or events, that are associated with some past trauma they have encountered. Similarly, we may become sentimentally attached to objects or events that remind us of good times in our lives.

Although behaviourism developed from Pavlov's discovery of a mechanism that could explain how fears, phobias and attractions might develop in both animals and people, it soon became the basis of a scientific movement that saw conditioned responses as the basis for all of human consciousness. Indeed, for a strict behaviourist, the very concept of consciousness made no sense in scientific terms. As it was impossible to ask an animal what was going on within its mind, and even human introspection could present a false or distorted view of what was going on inside a person's head, it was far better to treat the mind as an unknowable 'black box' and focus on what could be measured, that is behaviour.

At the same time, based on the discovery that animals could be taught all sorts of tricks using conditioning and a reward system, the idea grew within this movement that behaviour was infinitely malleable. All that mattered was environment, and if that changed, then so would the behaviour. The potential social implications of this viewpoint were expressed by the behaviourist John Watson when he stated: 'Give me a dozen healthy infants, well-formed, and my own specified world to bring them up in and I'll guarantee to take any one at random and train him to become any type of specialist I might select – doctor, lawyer, artist, merchant-chief and, yes, even beggar-man and thief, regardless of his talents, penchants, tendencies, abilities, vocations, and race of his ancestors.'[5]

In many ways, behaviourism echoed the 'blank slate' view of the mind pioneered by John Locke and David Hume in the 18th century, which I have already mentioned. However, behaviourists claimed that, unlike previous speculation, their viewpoint was based on the latest scientific analysis. Soon behaviourism became the dominant trend in psychology, in countries as apparently different as Stalinist Russia, where it was used to justify top-down social engineering, and post-war USA, which espoused the ideal of the free market and promised opportunities for everyone to progress in society if they worked hard enough.

Behaviourist flaws

From quite early on, however, it became clear that there were major flaws in behaviourism's claim that the brain could be treated like a black box. For instance, far from conditioning experiments in animals showing that their behaviour is infinitely malleable, it soon became clear that a particular species' ability to learn new forms of behaviour is much more constrained than had been thought, and that behaviours that could be learned were often closely linked to the types of activities the species would normally perform in the wild.[6]

Flaws also began to emerge in behaviourism's accounts of human behaviour and society. In the 1960s, the linguist Noam Chomsky took particular exception to a claim by the leading behaviourist, B.F. Skinner, that children learning to speak their native tongue was as much a conditioned response as any other form of human behaviour. According to Skinner, language development in children was simply a matter of a child hearing a word and beginning to associate it with an object or activity in the outside world.[7] Adults can reinforce such learning by praising a child if they make a correct word association or by correcting them if they don't. Yet Chomsky showed that this view of language acquisition was inadequate. Drawing on his linguistic insights, he showed that there is far more to human language than simply learning to associate words with objects and activities.

Instead, as I have already mentioned, a key aspect of language is its grammatical structure, which allows expression of complex notions of time and place. As Chomsky noted, an astonishing feature of our species is how rapidly and seemingly spontaneously children learn vocabulary and also how to structure words in a grammatically correct manner. In contrast, as I mentioned earlier, attempts to teach sign language to apes have shown that although some apes may learn to associate a considerable number of words with objects and even with emotional responses, there is no evidence that they can learn grammatical rules, or indeed have any ability to express time, space or any complex concepts in the way that human beings do when communicating.

To explain these differences between human beings and other primates, Chomsky proposed that humans have a biological 'language instinct' that other species lack.[8] However, subsequent studies have suggested that human language acquisition involves a far more complex interaction between a developing child's brain and its social environment than we get from either the behaviourist view or Chomsky's idea of a language instinct.

For instance, recently it has been proposed that oxytocin – a hormone mentioned previously that mediates social interactions – may be involved in language acquisition. This is based on recent studies that indicate that oxytocin plays an important role in the process by which songbirds learn to sing. As there are interesting parallels between this process and language acquisition in human children, the neuroscientist Erich Jarvis has proposed that 'when a child says "Daddy" and receives a pat on the back or a smile, that gives the child a feeling of reward', and suggested that such a 'feeling may release oxytocin into the vocal-learning circuits, to strengthen the memory in the vocal-learning pathway of how to say that sound'.[9] If this is true, it would challenge the idea that we simply learn to speak because we have a language instinct and show that human interactions are also critical to language acquisition. Since defects in oxytocin release have also been linked to some forms of autism, this might explain why some autistic individuals have language abnormalities.

Despite such new findings challenging an overly simplistic idea of a language instinct, what Chomsky did succeed in doing in the 1960s was to reassert the importance of the human brain as a factor in understanding human consciousness. However, given we now know a lot more about the molecular and cellular mechanisms that underlie brain function, how can we link this to what I have said about the importance of language for human consciousness?

The challenge here will be to link the emerging new information from neuroscience about brain mechanisms to what I have said about the unique aspects of humans as a species, to provide a materialist explanation for human consciousness, yet without falling into the trap of seeing this as guided by an inner homunculus. However, that will also mean challenging some philosophical assumptions within neuroscience, which is what I am now going to do.

INFORMATION AND MEANING 11

Philosophical ideas may influence the study of human thought in both direct and indirect ways. We previously looked at some of the viewpoints held by different philosophers of mind. Because philosophy as a discipline prides itself on being a rational and logical way of understanding the world, including our own minds, it tends to be explicit in its arguments. In contrast, an individual may espouse a particular philosophy without realising they are doing so. This is something we should be aware of in neuroscience, for even if its practitioners state that they are merely expressing 'facts' about brain function, they may still be implicitly influenced by one or other philosophical viewpoint, even if they profess to have no interest in philosophy as such, and this can have an important influence on their interpretation of their findings. Undoubtedly the biggest philosophical influence on neuroscience over the past century, and one that continues to play a dominant role, is behaviourism.

I mentioned previously that a significant backlash occurred in the 1960s against behaviourism's 'black box' approach to the mind. Chomsky criticised the idea that language acquisition in humans was merely a form of conditioned learning, but this was also part of a general growing criticism of the idea that human behaviour and society were really analogous to rats pressing levers for a reward. Yet despite this, behaviourism is still a very dominant philosophical influence among psychologists and neuroscientists, and it holds this influence through its transformation into a viewpoint called 'cognitive behaviourism'.[1]

Unlike classical behaviourism, which saw cognition as just a matter of stimulus and response, behaviour as malleable as the environment and the mind as an unknowable 'black box', cognitive behaviourism views the mind as a complex information processing device.[2] Essentially the human mind is seen as analogous to a very sophisticated computer that receives 'information' from the environment, processes this, and generates a response. So how valid is such an analogy? To answer that, we need to look more closely at how computers work and then examine whether brains work in an analogous way.

Computers nowadays occupy such a central part of our lives, and are involved in performing so many different roles in society, that it is easy to forget their basic function, which is to carry out sequences of arithmetic or logical operations. Although we think of computers as modern inventions, they have in fact been around for millennia, in the form of simple calculating machines such as the bead abacus. However, while the abacus can only be used for arithmetic, a modern computer is a general-purpose machine that can perform many tasks and does so at such a rapid rate that it has become a revolutionary technology.

Computing power

What all computers have in common, though, from a simple abacus to one of the supercomputers used by the US government, is that information is fed into them; this information is then processed by means of rule-based, written operations also known as algorithms, and then the processed information is received by the computer's operator. Today's electronic computers are also digital, that is they manipulate and store data in the form of ones or zeros, as these are easily represented with simple on–off electrical states.

Because computers only process data in an unconscious fashion, any 'information' that is fed into a computer remains meaningless to it, as is the final output that it generates.

If this is how a computer operates, how does a brain compare? One potential similarity between a brain and a computer is that both involve circuits. In the case of computers, these may be actual physical circuits or, more commonly nowadays, ones simulated by the computer's software. Yet here we have to be careful not to be misled by superficial similarities. Despite some types of computer circuits being named 'artificial neural networks' because their circuitry is supposed to be based on that of the brain, a crucial difference between these and biological neural circuits is that the latter's neurons are not simple on–off switches but complex biological entities.[3] In addition, as I have already noted, far from brains only consisting of electrical circuits of neurons, glial cells also contribute in fundamental ways to brain function, and as I will be exploring in more detail later, there is exciting new evidence from recent studies indicating that brain waves of different frequencies mediate important interactions within and between brain regions.

Linked to such differences between brains and computers, there are fundamental differences in how these two entities process information. As I noted above, human operators input meaningful information into a computer and the computer then transforms this information by subjecting it to various programmed algorithms. This hopefully then produces a meaningful, useful answer. However, as I have also noted, computers have no sense of the meaning of the data that is fed into them or the output data received by their operator.

In contrast, a human brain receives many incoming signals from the environment that in themselves are not meaningful information, yet the brain is then able to extract meaning from them. So how is this possible? And are the same, or different, forms of extraction of meaning taking place in a human brain, compared to that of an animal?

In considering how meaning might take shape within the brain, a place to start would be to look at how evolution has shaped brain function. I believe this is an important place to begin, because a fundamental difference between the human brain and a computer is that while the computer is a human-made object, the brain is a product of millions of years of biological natural selection. This means that whatever we discover about the role of the brain in the formation of human consciousness, we can be sure that it has been biologically selected as an object that will fulfil the survival and reproduction of its human owner, and that the different parts of the brain have evolved to work together in a successful fashion, in ways influenced by that selective pressure, rather than any sense of how a brain 'ought to work', which might be true if brains were designed like a computer.

Manipulating meaning

A second point to consider, given that a central claim of this book is that human consciousness has been fundamentally transformed through the twin media of tools and language, is how meaning might be created in the human brain in a self-conscious manner that is both uniquely different to the 'meaning' created in animal brains, while being based on the same, unconscious biological mechanisms that operate in all brains, whether animal or human. So now having considered these two points, let us now explore how they apply to sensory perception in the brain and how this might differ between an animal and a human.

We have already seen how sensory perception by the brain involves a number of different processes. First, some feature of the natural world must be registered, whether a specific wavelength of light for visual perception, a single chemical molecule for detection of odours or the equivalent for hearing, touch and taste. In fact, sensory perception involves far more than just the standard five senses, since our brains are also receiving signals from our inner ear, which informs our brain about our posture and balance, and our inner organs.

Given that all of these signals can contain no pre-programmed 'information' as such, how do our brains interpret these incoming signals and extract meaning from them? One place to start would be to think about the nature of the raw sensory data that enters our bodies and how this is interpreted. I have already mentioned that even in the first unicellular organisms, 'receptor' proteins had evolved to specifically respond to a particular wavelength of light, type of chemical molecule and so on. Therefore, if we want to think about incoming data in terms of 'information', it is worth pointing out that even at this most 'basic' level of sensory perception, some form of discrimination is operating.

Such discrimination is important in terms of understanding differences in perception between species and also between individuals within a species. So, for instance, when a bee looks at a sunflower it sees intricate patterns that are only visible in the ultraviolet part of the light spectrum, whereas when humans look at sunflowers, we see no such patterns.[4] Patterns on flowers that are only visible in the UV spectrum have evolved to strengthen the link between bees and certain plants in a way that helps both organisms, since the UV pattern acts as a signal to direct bees to the plant's pollen. However, it also means that in this respect, bees and humans are radically different in how we view the world.

One thing that bees and humans do have in common is that particular stimuli received by our sensory organs stimulate specific receptor proteins in those organs. This means that even at the most basic level of sensory perception, different stimuli activate distinct molecules in the body. This seems relevant in terms of one of the challenges posed by the 'hard' problem of consciousness, which is to explain that uniquely subjective aspect of human consciousness, for one thing we can say is that the distinctive sensory experiences of any individual involve a molecular, biological element. There is also the possibility that molecular biology may underlie subtle, subjective differences between human beings.

Take, for instance, red–green colour blindness, which is a relatively common human condition.[5] It generally only occurs in men, being due to a mutation in a gene on the X chromosome. Since women have two X chromosomes, they are usually protected from any effects of a mutation in an X chromosomal gene, as they will almost always also have a normal version of such a gene. However, with only one X chromosome, men lack this protection. Because men with red–green colour blindness grow up with this defect, their vision seems perfectly natural to them, and it is only through tests conducted during childhood that their different type of vision is revealed to them as abnormal. Based on emerging evidence about the genes involved in our ability to smell, taste or hear different sensory stimuli, it seems likely that subtle variations that occur in these genes in different people will mean that every individual may have their own unique perception of the sensory world and we should at least be able to partially relate to their specific genome.

Subjective biology

If, as part of tackling the 'hard' problem of consciousness, we ask how can we ever explain in material terms what it feels like to see a harvest moon, taste a ripe peach or experience Mozart's Queen of the Night aria from *The Magic Flute*, we should be open to the possibility that such subjective experiences might literally differ between individuals, in ways that are due to genetic variations in their sense organs' receptors.

In fact, as well as such genetic variations affecting human sensory perception, there is also emerging evidence that such perception can be affected in subtle ways by life events that change our sense organs. I am thinking particularly here of the way that the SARS-CoV2 virus affected some people's ability to smell and taste during the Covid pandemic.

Quite early on in this pandemic it became apparent that some people who had become infected by the SARS-CoV2 virus lost their sense of smell and taste, sometimes for considerable periods of time after recovering from the infection. Even when these senses returned, often they were now disturbingly distorted. For instance, a Parisian woman Hélène Barre reported that peanuts now smelled like shrimp, raw ham like butter and rice like Nutella.[6] While disconcerting for any individual, this was particularly a problem for Barre since she was an oenologist, an expert on wines, whose daily job was based on being able to detect the subtle differences in taste and odour that distinguish a great wine from a mediocre one. This has also been a problem for chefs, sommeliers and perfumers.

We do not fully understand how exactly SARS-CoV2 affects smell and taste in such an idiosyncratic fashion in some individuals, but there is evidence that the virus can directly affect the neurons in the nose that constitute the first stage in sending odorant signals to the brain. This shows that not only genetics, but other biological changes in our sense organs, can have quite profound effects on the ways that we perceive the world around us.

As already mentioned, the initial harvesting of sensory signals by the sense organs and related sensory structures is only the first step in the process of perception. Equally important is what then happens in the brain, yet this is the part of the process that remains far from clear. In particular, we are only just beginning to gain a better scientific understanding of how perception combines with the rest of consciousness to give us that sense of being a unique individual human being, soaking up all that the world can throw at us, feeling it in our own individual way and then relating this to all the other facets of our personality. And it is time now to move forward from thinking about how the raw sensory inputs are collected and turn again to how those inputs take on meaning within the brain. As part of doing that, I now want to consider again how evolution has shaped the human brain and the ways that this might have affected brain function, compared to a designed object.

CHANCE AND DESIGN 12

Charles Darwin and Alfred Russel Wallace precipitated a revolution in our understanding of the natural world when in 1858 they independently proposed that all life on Earth came into being through an evolutionary process they called natural selection. Their great joint insight was to recognise that, within any biological species, there will be subtly different variants, and because some variants have specific characteristics that make them better able to survive in a particular environment, they will be more likely to produce offspring that also have these characteristics. Eventually, more and more individuals within a species will have the favourable characteristics and, over time, this can change the character of a species and even lead to new species arising if diverging variants are separated, say by geography. For instance, because finches of one original species colonised the different islands of the Galapagos, off the coast of Ecuador, and since these islands had different types of food sources, finches on the different islands evolved different types of beaks. Indeed, it was observation of such finches that helped guide Darwin to his great insight.

One of the most revolutionary aspects of the theory of natural selection was the proposal that evolution was purely driven by blind chance. This was also one of the most threatening aspects of the theory for people who had been brought up to believe that the lifeforms on Earth were all part of God's design, and as such it directly challenged some religious viewpoints. It was also Darwin's insistence on the fundamental role of blind chance for all evolution that distinguished him from Wallace.[1] Wallace found it impossible to accept that something as complex as the human brain was a product of chance, and instead argued that it must have been created by 'a Power which has guided the action of [natural] laws in definite directions and for special ends', in other words maintaining the idea of a supernatural creator. Darwin was aghast at this; he felt evolution only worked if it applied to everything and that making an exception for the brain could threaten the whole theory. It was this fear that led Darwin to write to Wallace: 'I hope you have not murdered too completely your own & my child,' in reference to the principle they had co-discovered.

Any materialist theory of human consciousness put forward today must surely follow Darwin in seeing the human mind as ultimately a product of the same blind chance that has guided the evolution of all other biological entities on Earth. Yet it is one thing to state this, another to explain the 'hard' problem of consciousness in such a way, and in this chapter I would like to explore in more detail the consequences of adopting such a guiding principle.

Unconscious design

One important consequence of seeing the human brain as purely a product of natural selection is that we have to reject any idea that the brain is the result of conscious design. This may seem obvious, but it is important to note, given what I said previously about cognitive behaviourism's use of the computer as a model of human brain function. For, unlike a brain, all computers are created by a designer, and their structure reflects this, not just in their hardware but in the hierarchical way their components work as a whole.

Typically, a computer has input devices, such as a mouse and keyboard, or touchpad, and output ones, such as video monitors or audio speakers. It also has a memory, generally divided into short- and long-term storage and a central processing unit, or CPU, which is the electronic circuitry that executes instructions comprising a computer program.

I have already mentioned one problem with a computer analogy for the mind, which is that meaningful information is fed into computers, which unconsciously process it and then transform it: hopefully still meaningful information is then interpreted by the computer's human operator. In contrast, meaning is somehow created within a brain, since what enters it is raw data, but what emerges is something relevant for the brain's owner. Somehow we need to explain how such a process can occur in the biological brain.

I now want to note another problem with the computer analogy, which is the fact that, unlike a computer, the brain surely cannot have a central processing unit, for if it did this would imply that it had some controlling program, an idea that we have to reject as this would require an inner homunculus to have designed such a program. Yet if we reject such an idea, does this mean that we also have to be cautious of claims that there are 'higher' centres of brain function? It is now time to look at both these issues.

To start with creation of meaning within the brain, let us think in more detail of what happens to incoming signals as they enter the brain and how their transformation into something meaningful to the brain's owner might occur within a purely biologically based mind. I have deliberately used the word 'signal' here instead of 'information', seeing it as a less loaded term, for while signal merely refers to any entity that is transmitted, information implies some kind of content that is meaningful to the individual who receives it. Yet one thing we can say about signals is that they do have a specific structure and a form that may carry information if there is some way of meaningfully interpreting it within the brain. I would now like to look at how this might be possible through the process of 'cell signalling'.

Cellular signals

Cell signalling is a central process for lifeforms ranging from simple bacteria to humans.[2] I mentioned earlier that bacteria respond to chemicals in the environment through receptors on their surface, but such receptors are only the first stage of a chain of events within the cell. Typically, a receptor will trigger activation of another protein that will then activate a further protein and so on. A distinctive characteristic of signalling pathways is that they are very varied in form, but extremely specific in terms of the precise sequence of proteins activated by a particular receptor. In terms of sensory perception by the brain, this means that a particular chemical molecule in food detected by the tongue, or an odorant molecule detected by the nose, will trigger a signalling pathway in sensory neurons that is highly specific for that molecule. In the same way, the rods and cones of the eye that respond to light of different intensities and wavelengths or the cells of the inner ear that respond to sounds of different amplitudes and frequencies also produce a signalling response that is highly specific to some pattern in the visual or auditory environment.

If precise and highly varied signalling pathways are the way that our sensory organs record the molecular content of the environment, also very important is the architecture of those organs and how this architectural structure has a parallel in brain structure.

I have already mentioned how this works in terms of vision. Neurons in the visual cortex are organised into interconnected columns and layers whose combined activity helps to recreate the complex make-up of the outside world within the neural architecture of the brain. In a similar fashion, the structure of the ear, nose, tongue and the patterning of the touch receptors on the skin are linked to sensory neurons in such a way that signals entering the brain reflect not only the molecular content of the outside world, but the source of a scent or a sound and the specific part of our skin where we have touched something or been touched. So how does all this register as meaning in the brain and how is any of this related to the points I made earlier about natural selection?

One way in which natural selection has shaped sensory perception is in the evolutionary development of the receptors and linked signalling pathways involved in such perception. In multicellular organisms like ourselves, there has been a dramatic expansion and diversification of signalling pathways since our ancestors evolved from unicellular life. This has allowed such organisms to not only perceive a much greater diversity of chemicals in the environment but also to register changes in their own internal bodily chemicals; adrenaline is one such chemical and another one is oxytocin, the former being central to the response to danger, the latter underlying an emotion such as love.

One issue to consider in this context is whether there is any necessary connection between the chemicals in our bodies and the emotions associated with them. What we do know is that adrenaline, or chemicals almost identical to it, trigger the same 'fright, fight or flight' response, not just in mammals and other vertebrates, but also in many invertebrate species.[3] This tells us that adrenaline's link with this response is an ancient one. However, it is not clear why this chemical in particular took on this role as a signalling molecule, and given that evolution is driven by blind chance, it may well be a coincidental association.

Adrenaline is not a neurotransmitter but a hormone that is released into the blood by the adrenal gland. A related chemical, noradrenaline, plays a similar excitatory role in neurons in the brain and also ones that connect with the heart and many other organs.[4] Another related chemical, dopamine, which I have mentioned previously, is an important neurotransmitter in the brain, where it acts to activate and stimulate brain regions. Thus, while a chance incident may have led to adrenaline-like chemicals first beginning to mediate the response to danger long ago in evolutionary history, it is no accident that chemicals of this type are now associated with attention, alertness and readiness for action, both in the body and the brain, and we can thank natural selection for the initial link and its later evolution.

Natural selection is also the central force that has guided the evolution of the human brain. Darwin recognised that one of the biggest challenges his theory faced was explaining how a complex organ can evolve purely through blind chance. In addressing this challenge he was acutely aware of an argument made by the theologian William Paley in 1802, over 50 years before the publication of *The Origin of Species*.[5] Paley's argument was written in the form of a story about a person walking along a moorland path, who comes across a watch, an object they have never seen before. The person wonders whether the object has come into being by accident or been designed by someone, and eventually comes to the conclusion that it must have been designed, because of the exquisite and interconnected way in which the watch's individual parts are put together. Paley then turns to a biological object, the human eye, and argues that because the eye is also clearly an object in which all the individual parts are assembled together in a complex and sophisticated fashion, this proves that it cannot have evolved by chance but must have been designed by God, as must the rest of the human body, and indeed all the other lifeforms on our planet.

Creative chances

Darwin took Paley's arguments very seriously. He agreed that it was difficult to imagine how an object as complex as the human eye could have come into being through chance, but only if we did not consider all the simpler evolutionary forms that preceded it. By gathering information about the visual systems of a variety of organisms, he showed there is a whole spectrum ranging from a simple patch of light-sensitive cells in limpets, to a more complex water-filled cavity with light-sensitive cells at its base in a type of marine snail, to the highly complex camera-like eye with a lens and focusing muscles in a human being. In other words, if we look at such a range of species, instead of just focusing on the most complex ones like ourselves, we can clearly see how the eye has evolved from very simple beginnings to the highly complex object that we use to view the world around us.

In fact, a similar type of argument can be made for other complex organs. The human heart, with its four interconnected chambers, pacemakers that regulate its pumping action and its connection to the lungs and the rest of the body through arteries and veins, is so well formed that it may seem hard at first to see how it could have evolved through chance. Yet a survey of hearts and circulatory systems in other species reveals a variety of structures ranging from a simple contractile tube in flies, to a simple one-chambered heart in fish, to three chambers in a frog's heart, to the complex four-chambered object in humans.[6] Similar evolutionary 'stages' can be found for other organs like the kidney, liver and lungs.

As we have seen, the origins of the brain can be traced back to the receptors that unicellular species use to interact with their environment, and the stimulus-response neural networks of simple invertebrates like jellyfish. Then multicellular organisms became increasingly complex, by the need to respond not only to the external world but also coordinate the activities of the different types of cells, tissues and organs. To illustrate brain evolution, a survey of brains from organisms ranging from flies to humans reveals a clear increase in complexity. Even within mammals, a comparison of a mouse and monkey brain shows that the latter is much more complex and I have already mentioned the dramatic relative increase in brain size that occurred during the evolution of humans from apes. However, while such comparisons show how evolution has changed the brain in terms of gross structure, we face a major challenge in understanding how natural selection has shaped the brain at the level of finer structural detail, or indeed how evolution has affected the way that the brain is interconnected and how this relates to its function. This is related to a point I made previously, which is that for all the advances in neuroscience in recent decades, it is still far from clear how the brain functions as a whole.

Given this caveat, from what we do know about brain structure and function, what can we say about the likely role of natural selection in its development and what further insights might we try and gain based on the model of consciousness proposed in this book? To pursue this matter further, I now want to look at the brain as a whole entity and try and connect that to what I have said about the creation of meaning within the brain.

STRUCTURE AND FUNCTION

13

Of all the influences on architecture in the 20th century, the Bauhaus School has been one of the most important.[1] Founded by Walter Gropius in Weimar, Germany, in 1919, its ethos that structure and function should be intimately related continues to be an important influence today. In the human body, structure and function are also intimately related, not because anyone designed our organs, cells and the molecular structures within them, but through the power of natural selection. Because this operates by elimination of bodily and cellular components that do not benefit survival of the organism, it can be ruthlessly precise in favouring components that work well and eliminating those that do not, and over millions of years this can lead to the appearance of planned and often exquisite design.

I have already mentioned that the heart is a good example of an organ with a clear link between form and function. Crucially, William Harvey recognised that the heart functions to pump blood in a one-way direction through the circulatory system, so that oxygen can be carried to the tissues and organs that require it and waste carbon dioxide removed.

It is testament to the importance of machines in human society that we tend to use mechanical analogies to describe the different organs of the body. Therefore, the heart is a pump, the eye a camera, the kidney a filter, the liver a detoxifier and so on. Such analogies can be very useful in allowing us to understand how organs work in common-sense terms, but they can also be misleading in giving the impression that organs only have one function. So while it is true that a central function of the heart is to pump blood, we are only more recently beginning to recognise its important role in secreting hormones into the blood that target the kidney and thereby regulate blood volume, pressure and salt concentration.[2]

What about the brain? I have already mentioned some of the problems with using the computer as a model of brain function. Therefore, if we reject this analogy, can we find a different type of unifying model that can explain brain function, as Harvey did for the heart?

One potential problem in searching for such a model is the complexity of the brain. Yet for all such complexity, as Darwin recognised, the human brain is as much a product of natural selection as the heart, and there is no reason to believe that its form is any less related to function than is the case for the heart, so brain structure is a good place to start.

Having already raised the question of how meaning might be created from incoming raw data in the brain and explored some of the ways that this might work in terms of sense organs and signalling pathways, I now want to look at this issue, but with regard to the different regions of the brain, their structure and how they might interconnect.

We have already seen how potentially meaningful information enters the brain both because of the specific receptors in sensory organs and the way signalling specificity is maintained through subsequent signalling pathways. What happens next in this process?

Visual architecture

To take vision as an example, we saw how the architecture of the visual cortex to an extent recreates the shapes, textures and other characteristics of objects in the outside world. This connection between brain architecture and the different features of objects in that world must have evolved through natural selection, such selection favouring a situation in which the brain could accurately represent the world around us in a precise and ordered fashion. Yet in itself the cortex has no awareness of the objects that stimulate it. Visual awareness in an individual can only come about through further connections of the visual cortex with other regions of the brain. There is also a reason to believe that such connections arise both from evolved brain structure and the impact of the environment.

To explain further what I mean, consider face recognition. Studies have shown that human babies show a particular interest in faces or face-like objects as early as two days old.[3] Presumably, such is the importance for survival of being able to recognise a carer's face that natural selection has led to such a preference for face-like objects to be hardwired in the brain of a newborn child. Later, at six months old, babies can distinguish a familiar and unfamiliar face, showing that learning and memory have also become important components of face recognition. Indeed, over the course of their life, a typical human being may come across hundreds of thousands of different faces, and it is one of our special skills as a species that we can recognise the face of an old friend or former colleague, say on a crowded city street, among many other people and having not seen that individual for years.

In fact, we share the ability to recognise one face among many with other primate species, reflecting the importance for such a highly social animal group as primates in being able to distinguish individual members of their species. Yet we have also seen how studies of people undergoing brain surgery showed that a specific neuron in the brain could become stimulated by a photo of the actor Halle Berry, but also by a cartoon of her, her face obscured by a mask when she was playing 'Catwoman' in the film of the same name and even by a picture of her written name 'Halle Berry'. This suggests that in human beings the image of a person can somehow become linked in the neuronal circuitry of the brain to objects that are related in much more indirect ways to the concept of that person.

In fact, this must be equally true of our other senses. The noise of an aeroplane flying overhead is something that would have been foreign to human beings only a century and a half ago, as the first aeroplane was only flown in 1903, by the Wright brothers. Yet it is such a familiar sound now that we recognise it instantly and know what is causing it, without even looking up into the sky, because we have a general concept of aeroplanes as metal objects that make a noise and fly. Meanwhile, hearing a memorable piece of film music may remind us not only of the film in which that music appeared, but a specific scene linked to the music and even our previous personal emotional response on watching that scene.

Unlocking memories

Similarly, a particular scent, taste or even the sensation of something familiar on our skin will not only create a link to a memory of that specific sensation but potentially also to other memories that are only indirectly related to that particular memory. Famously, in Marcel Proust's novel *In Search of Lost Time*, the taste of a madeleine cake dipped in tea is enough to trigger an involuntary series of long-forgotten memories about the narrator's past life.[4]

Of course, we are not the only species to have complex memories. For instance, I would imagine that when my cat Maya goes out into our garden, not only will her highly sensitive nose pick up all sorts of familiar scents but such scents will trigger all sorts of interconnected memories about the animals, plants and other objects that are the source of those scents and may extend some way into her life history. Yet as I have argued previously, what will be lacking in a cat's mind is a language-based conceptual framework.

We tend to think of memories as relevant only to things that happened in the past, with a sensual experience acting as a trigger to remind us of a past event or indeed an interconnected series of past events. Yet recent studies have challenged this viewpoint. Instead, they have revealed striking similarities in the brain processes involved in remembering the past and those that help us to imagine or simulate the future.

For instance, a study in which volunteers were asked to remember past events or imagine things that might happen to them in the future, while their brains were being imaged, showed that similar parts of the brain became active during the two types of task.[5] Studies of individuals who have lost their short-term memory due to damage to their hippocampus, like Henry Molaison, also tend to have problems imagining novel things that might happen to them in the future.

Molaison himself, when asked what he believed would happen tomorrow, tended to answer 'whatever is beneficial',[6] but when asked to make a prediction about his personal future, he sometimes picked out an event from a distant memory, but often he did not respond at all. It seemed that Molaison had no stored information available to allow him to weigh up the possibilities available to him in the future. Of course, this may simply reflect the fact that people tend to reach into the experiences of their past when trying to imagine how they might respond to a particular situation in the future, but the findings suggest that we need to be careful when separating brain functions into distinct categories.

In fact, a number of recent studies have begun to challenge the idea that aspects of consciousness that we have tended to distinguish using terms like perception, attention, motivation, memory, imagination, reason and so on, are really separate in the brain. And I have already said that I think we need to be cautious about the way we use terms like 'information' or 'higher' centres to refer to brain processes. With that in mind, can we find evidence from the latest neuroscience research that can do justice to the complexity of human consciousness, but also be based on real, biological brain mechanisms? I believe we can, but to do so, we need to look more closely at brain architecture and how this influences the way a human brain works, in other words, how structure relates to function.

In linking structure and function in the brain, we need a clear idea about what its different components do. To some extent we've succeeded in this goal. There is therefore good evidence that the hypothalamus, a tiny structure deep in the brain, regulates processes of major importance to our existence, like body temperature, appetite and thirst, the body clock and sexual desire. Such is the importance of this structure that if it is destroyed during an injury to the brain, usually, the affected individual will be unable to survive.

The hypothalamus is not the only brain structure whose role we now understand much better because of the effects of its loss through injury. Such injuries may occur in a human being through botched surgery, in a traffic accident or as a result of crime or war. For instance, we saw that the hippocampus was only recognised as playing an important role in memory following the horrific impact on this faculty when it was removed from Henry Molaison as part of an attempt to stop the epileptic seizures that he was experiencing.

In addition, selective destruction of certain brain regions in experimental animals is one of the approaches commonly used by neuroscientists to study their role in brain function. Such studies have been used, for instance, to confirm the importance of the hippocampus for memory, to study the role of the cerebellum as a regulator of repetitive movements and to investigate the involvement of the amygdala in the fear response.

The vast majority of studies of experimental animals in neuroscience involve mice and rats. One caveat of such experiments is that the rodent brain is not only much smaller than that of a human but it also differs significantly in structure. However, it is not only in terms of gross brain anatomy that humans and rodents differ; recent studies have also identified key differences between rodent and human brains at the molecular level. For this reason, studies of non-human primates' brains continue to have an important role in neuroscience.

I mentioned previously how humans evolved from apes when our ancestors began to walk on two legs, which freed the hands for using and designing tools and had such a transformative effect on proto-human society that it led to the development of language and also to dramatic changes in brain size and structure. Yet it is also important to recognise that such changes to the brain occurred in an organ that had already undergone significant changes during the evolution of the primates as a distinct animal group.

Prefontal planning

One particular brain region that is greatly expanded in human beings is the prefrontal cortex.[7] This brain region, which is located just under the forehead and accounts for about 10 per cent of brain volume, has acquired a special significance in discussions about the material basis of human consciousness. This is because this region is commonly associated with brain functions related to human uniqueness, such as reasoning, planning, decision making, control of social behaviour and some aspects of language.

Support for the idea that the prefrontal cortex is responsible for such 'executive' functions of the human brain has come partly from observation of unfortunate individuals who have suffered damage to the prefrontal cortex. The most famous such individual was a US railway foreman, Phineas Gage.[8] In 1848, Gage was using an iron rod to pack dynamite into a hole during construction of a railway when the explosive went off prematurely, sending the rod through the roof of his mouth and out via the top of his skull.

Amazingly, Gage survived the accident, and he could subsequently walk, talk and generally function as before. However, his behaviour was said to have subsequently changed dramatically; he was reported to have gone from being a highly motivated and organised individual to one who could no longer make plans and who had also lost all social grace and restraint. When Gage died some years later, an autopsy showed that one of the brain regions that had been particularly damaged was the prefrontal cortex.

In fact, the scale of the changes in Gage's personality are now somewhat disputed – after all, he did successfully hold down several jobs after his accident – but enough subsequent, detailed scientific analysis has been performed on individuals with more subtle damage to the prefrontal cortex to support the connection of this region with executive brain functions. Such analysis has shown that these individuals tend to perform poorly on tasks that require the use of long-term strategies and the inhibition of impulses, but they can also have short-term memory defects and blunted emotional responses.

Although the prefrontal cortex is particularly prominent in humans, it is also expanded in primates as a whole compared to other mammals, so it is worth considering why this evolutionary change occurred and what we might learn about the material basis of human consciousness by studying the prefrontal cortical function in non-human primates.

The question of why the primate brain, and particularly the prefrontal cortex, began to expand in size relative to other mammalian groups, is a matter of debate, but one plausible explanation is that it was linked to the growing complexity of primate society. For one characteristic feature of primates compared to other mammals is the fact that the former tend to live in relatively large, stable, bonded social groups. While this arrangement has many benefits, for instance, protection from predators, it also poses challenges. One of these is how to successfully coordinate the activities of a large group; another is how to minimise the tensions that may arise through such coexistence. It may be that this was the particular stimulus that led to the development of a brain region that could deal in a more sophisticated manner with large-scale group interactions. In line with this, the biologist Robin Dunbar has recently shown that overall brain size for a particular primate species correlates with the size of the social group for that species.[9]

A central question to now address is how the prefrontal cortex functions in such a way, both in its inner workings and how this relates to the rest of the brain. In fact, this has become a very exciting area of research in current neuroscience, for recent studies in primates have not only dramatically challenged the way we view prefrontal cortical function, but also form part of a growing awareness of the important role played by brain waves of different frequencies in connecting different brain regions in a highly dynamic way, and also controlling different aspects of consciousness, in a frequency dependent fashion.

CIRCUITS AND WAVES 14

It is a remarkable fact that all of human consciousness, from the most basic thought to the height of artistic or scientific creativity, is based on the same phenomenon – electricity – that illuminates a lightbulb, heats an electrical oven or powers the computer that I am using to write these words. For despite the almost unimaginable complexity of the human brain, with its estimated 86 billion neurons and 100 trillion neural connections, and despite what I have said about the importance of glial cells, the brain is basically an electrical device.

In fact, this feature of nervous systems, if not brains themselves, was first recognised as early as 1780, when the scientist Luigi Galvani and his wife Lucia Galeazzi Galvani discovered that the muscles of dead frogs' legs twitched when activated by an electrostatic spark.[1] Based on this observation and subsequent experiments, the Galvanis concluded that the motive force in the nervous system was electricity rather than some kind of 'vital force'. It was only a century later, however, that observations of brain tissue under a microscope by Ramón y Cajal, in 1888, finally showed exactly how electricity was transmitted in both the brain and the peripheral nervous system by identifying neurons as the units of transmission, and demonstrating that these were linked together in circuits.[2] As we have seen, we now know that electricity moves along the axon of a neuron in the form of an 'action potential'; this activates the release of neurochemicals into the synaptic gap at the end of the axon and this then stimulates an action potential in the next neuron in the chain.

A few years before Cajal's discovery, in 1875, Richard Caton had shown that electrical activity within the brain could be recorded by the simple measure of attaching electrodes to the skull of a rabbit or monkey.[3] Caton noted that there were distinct types of brain 'wave', each with a distinct frequency. Yet in contrast to Cajal's discovery, such was the scepticism about the validity of Caton's findings that they were almost totally ignored by the scientific community. When Hans Berger, one of the few people who took Caton's discovery seriously, eventually followed it up by recording brain waves in a human being in 1925, he was so doubtful about his own findings that he did not publish them for a further five years and there was considerable scepticism about the findings for a further decade.

Today, the measurement of brain waves using an electroencephalogram, or EEG, is a commonly employed and important approach in clinical psychiatry that is used to diagnose pathological conditions affecting the brain, including stroke, epilepsy, schizophrenia and post-traumatic vegetative state. It has also been known for some years that different human mental states are associated with different frequencies of brain waves. So alpha waves (9–14 hertz) are associated with a relaxed state of mind, beta waves (1–30 Hz) with an alert state, gamma waves (31–100 Hz) with problem solving and concentration, theta waves (4–8 Hz) with deep relaxation and delta waves (1–3 Hz) with deep, dreamless sleep.

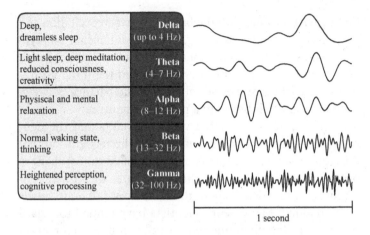

Deep, dreamless sleep	Delta (up to 4 Hz)
Light sleep, deep meditation, reduced consciousness, creativity	Theta (4–7 Hz)
Physiscal and mental relaxation	Alpha (8–12 Hz)
Normal waking state, thinking	Beta (13–32 Hz)
Heightened perception, cognitive processing	Gamma (32–100 Hz)

1 second

Figure 11. Different frequencies of electrical waves in the human brain.

Excited minds

We also now understand how brain waves of different frequencies arise. I mentioned earlier that the brain contains both excitatory and inhibitory neurons. There is a good evolutionary reason for this. If the brain contained only excitatory neurons, then there would be nothing to stop neuronal impulses sweeping across the brain in an uncontrolled manner. In fact, this is precisely what happens, to devastating effect, in a condition like epilepsy, where it can cause convulsive seizures.[4] Epilepsy, though, is a pathological condition. Normally, the fact that excitatory neurons have inputs from inhibitory neurons as well as excitatory ones prevents such an uncontrolled spread of neuronal impulses.

Another important point to mention is that any particular neuron will have inputs to its dendrites from the axons of many other neurons. Also, the fact that this relationship is a highly dynamic one, with the balance between excitation and inhibition constantly changing, means that neurons will typically show regular oscillations in their electrical potential. This then is the basis for brain waves and the different frequencies of these waves simply reflect the balance between excitation and inhibition in a particular neuron.

Yet despite the fact that we now understand the neuronal basis of brain waves, and despite the recognition that different wave frequencies seem to be linked to real changes occurring in the brain and different states of human consciousness, scepticism about the significance of brain waves has remained. Many neuroscientists have tended to regard them as an 'epiphenomenon', a reflection of inner brain activities, but not in themselves mediators of important brain processes, and certainly not fundamental for consciousness.[5]

That scepticism has begun to be challenged by some exciting new findings. A number of scientists have contributed to this awakened interest in brain waves, but I want to particularly focus on recent studies by two neuroscientists, Earl Miller and Pascal Fries.

Let us start with Miller's work. For several decades, his research has been focused on understanding 'working memory'.[6] In fact, this term is a bit of a misnomer. For although it can refer to a situation like the popular children's game 'pairs', in which players identify matching pairs of concealed cards by turning them over and remembering their identity and position, we use our working memory in many other ways in our daily lives. For instance, we can be guided by this type of thought process if following directions received from a friend about how to reach their new house or if we were in a shop selling different mobile phone brands and weighing up their positive and negative features before buying one. Consider a more complex task: waiting at an airport for a flight. To catch the plane on time we would need to keep checking the information screen, maybe while also having a restaurant meal or shopping, and then once details of the boarding gate appear, find the correct gate, where we would now need to remember where we put our passport and boarding pass. It is for this reason that Earl Miller refers to working memory as the 'sketchpad of conscious thought'.

A key aspect of working memory for Miller is its 'volitional' quality. As he puts it: 'It is the main mechanism by which your brain wrests control from the environment and puts it under its own control. Any simple creature can just react to the environment. But what higher order animals have evolved is the ability to take control over their own thoughts.'[7]

Given its proposed primary role in 'executive' or controlling aspects of consciousness, the prefrontal cortex should be a key mediator of working memory. However, it has to do so by coordinating the behaviour of other parts of the brain: memory, definitely, but also visual and other sensory inputs, and also brain regions responsible for motivation. It is through studying working memory in monkeys that Miller believes he has identified how such coordination between the prefrontal cortex and other regions takes place.

Memory games

For their studies, Miller and his colleagues trained monkeys to play various memory games in the simplest game, monkeys were shown an image and they then had to press a lever if, among a series of images, the first image appeared again. In a more complicated game, monkeys were shown a sequence of two images and they had to indicate when a following sequence of two had the same images in the same order. In a third type of game, monkeys had to discriminate between two classes of abstract image, these being a silhouette of either a cat or a dog. The monkeys were rewarded with juice if they picked the correct image or sequence of images. During the game, electrodes that had previously been implanted in different regions of the monkeys' brains recorded electrical activity.

A common theme in all the games was that the monkeys had to make some kind of value judgement in their choices, based on an ability to discriminate between different aspects of an image or sequence of images. It was thought likely that this would be a task for the prefrontal cortex given its 'executive' role, and Miller's original aim was to identify the specific neurons in this brain region involved in such discrimination and judgement.

A totally unexpected outcome of the studies was that around 30 to 40 per cent of neurons in the prefrontal cortex showed task-related activity during the games.[8] At first this seemed to make little sense, for a common view of how the brain worked at that time saw each neuron as having one specific task. Yet if almost half of the neurons in the prefrontal cortex were engaged in one task, surely that would mean this brain region would only be capable of two or three tasks, something that is clearly not the case in either monkeys or human beings. In the end, Miller realised that the only solution was to propose a different model in which, rather than each neuron having a dedicated, specific role, instead neurons in this region must be able to multitask. Moreover, they would do so by forming groupings, known as ensembles, with other multitasking neurons, on a temporary basis.

Such a viewpoint was so radically different from the traditional one that it unsurprisingly drew quite a lot of criticism. For instance, one leading neuroscientist accused Miller of 'turning the cortex into a bowl of porridge'.[9] Indeed, a potential problem with the idea of multitasking neurons was how would it be possible for their activity to be distinguished within the brain and mental tasks to remain separate? The answer came in the form of a change in thinking about brain function that occurred at this time – around the turn of the millennium – based around new findings. This was the recognition that the activities of neurons can become synchronised when such neurons become sensitised to oscillating electrical waves of the same, specific frequency. Pascal Fries, who has been an important pioneer of studies that led to this change in viewpoint, calls such synchronisation 'communication through coherence'.[10] It was through reading about this phenomenon that Earl Miller wondered if it could explain his findings.

Sure enough, when Miller's team began to investigate further, they found that working memory involves the coordination of different ensembles of neurons by brain waves of different frequencies that fulfil different roles in such coordination.[11] Essentially, higher frequency gamma waves coordinate ensembles of neurons that carry signals relating to incoming data about a memory of an object or a number. In contrast, lower frequency beta waves coordinate neuronal ensembles that carry signals from the prefrontal cortex that can both stimulate and select such signals, but also suppress them when a shift in direction is required. Miller believes the findings support the idea that gamma waves coordinate neurons carrying 'bottom-up' information and beta waves coordinate neurons that convey 'top-down' information, such as prior knowledge and goals. In line with this, gamma waves were detected in the uppermost layers of the brain cortex, where incoming sensory neurons are located, and beta waves were detected in deeper layers, where neurons carrying signals from the prefrontal cortex to the rest of the brain are to be found.

Another important finding made by Miller and his team was that ensembles differed in having distinct frequencies in the beta or gamma range compared to other ensembles in that range. He also proposed that this could allow different aspects of working memory to be held as distinct entities within the brain. Therefore, discrete 'packets' of memorised information could be held separate from each other by differential gamma labelling and different aspects of executive control could be distinguished by beta labels.

Such separation could be relevant to a scenario I mentioned earlier: waiting at an airport for a flight. In such a situation, I am sure that I am not the only person who often finds it a juggling act to keep checking the information screen while, say, enjoying a meal or shopping, then once the boarding gate details appear on the screen, keeping this in my head while I am walking to the gate and finally when I get to the gate, remembering where in my hand luggage I put my passport and boarding pass. Yet such a complex juggling act is routinely managed by millions of people at airports every day and the new findings suggest that it is made possible by such frequency-based distinguishing of neuronal ensembles.

Conscious waves

Perhaps the most exciting aspect of these new findings is that there seems good reason to believe that they are not only relevant to working memory but many other aspects of brain function. Indeed, Pascal Fries has recently presented evidence for a general role for such a mechanism in brain processes as diverse as vision, hearing, attention and movement, in species ranging from cats to monkeys to humans.[12] As well as confirming the role of gamma waves as coordinators of sensory input and beta waves as 'top-down' regulators of such input by 'higher' brain regions, he has also shown that theta waves seem to play a particular role in shifting attention, the process by which our minds stop focusing on one feature of the world and begin focusing on a different feature instead.

If the latest findings highlight the importance of brain waves in a variety of mental processes, where does this leave the more traditional model of cognition, based as this is on the idea of distinct roles for different brain regions consisting of, and linked by, precise circuits of neurons of different types connected together by synapses? It is important to note that the two ways of looking at the brain are not opposed, but highly complementary.

Understanding the precise architecture of the brain, be that of a mouse, monkey or human, remains key to understanding consciousness in these species, as does a detailed picture of the molecular and cellular processes that underlie how a neuron transmits signals along its length, how neurons interact and how glial cells assist in this process. However, the new findings introduce an important dynamic element. They show that since neuronal ensembles can be rapidly assembled and disassembled, in the process affecting the balance between different brain regions and 'bottom-up' versus 'top-down' signals, this provides a much-needed explanation for how thoughts can rapidly come and go in our consciousness.

With such new brain mechanisms now becoming apparent, how does this affect what we can say about the material basis of consciousness in both animals and humans? It is time to look at this question directly, and as part of doing so, I want to look again at the use of terms like 'top-down' or 'executive' control of brain function, which are commonly used in neuroscience, and see whether we can achieve a more precise understanding of how one part of the brain might exert control over the other, in terms of underlying mechanisms.

In fact, there are ways to explain how such a hierarchical situation might arise, that on the one hand relate to animal consciousness and on the other to consciousness in a human being. So first I want to look at the question of how natural selection might have created a hierarchy of brain regions and functions in certain animal groups, with a particular focus on non-human primates. Then I want to consider how the twin capacities for language and tool use and design have altered the brain still further in humans. Ultimately, the point of such an examination will be to not only provide a better understanding of how a human, or a monkey, brain works, but to take us once again to the centre of what it means to be a conscious, thinking human being. That means thinking about how such things as self-conscious awareness, a sense of free will, an ability to understand what others might be thinking and a capacity to reflect on the past, but also what might happen in the future, are related to brain mechanisms but also to our interactions with the outside world.

FREE WILL AND SELFHOOD

15

Of all the subjects debated within the philosophy of mind, psychology and neuroscience, the concepts of selfhood and free will must rank as among the most contentious. On the one hand, the sole thing that Descartes seemed to think we could be sure of when he stated that 'I think, therefore I am', was our own individual existence. For centuries, a linked widespread belief has been the idea that there is such a thing as free will. Indeed, the philosopher Immanuel Kant not only viewed free will as central to human existence, but he also saw it as a cornerstone of human ethics and law.[1] For example, if as individuals we are not free to choose, then it makes no sense to say we should choose the path of right versus wrong.

On the other hand, such certainties have recently been challenged. We have already seen how philosophers Daniel Dennett and Keith Frankish have argued that our feeling of a unified consciousness existing within our heads is an illusion. As Frankish himself expresses it: 'Illusionists do not deny the existence of consciousness, but they ... reject the view that it consists in private mental qualities and argue that it involves being related to the world through a web of informational sensitivities and reactions.'[2] Dennett thinks that this illusion of consciousness bears the same relation to our brains as the folders and other icons on the screen of a computer bear to its underlying circuitry and software.[3]

There has been a similar reassessment of the concept of free will. Psychologist Daniel Wegner has claimed that 'the experience of willing an act arises from interpreting one's thought as the cause of the act'.[4] In other words, our feeling of making a choice or decision is just an awareness of what the brain has already decided for us. A study often cited in support of this view was performed in the 1980s by psychologist Benjamin Libet using human volunteers. These were asked to perform a task like pressing a button or flexing their wrist, but also to note when they were consciously aware of their decision to move, while EEG electrodes attached to their head monitored their brains. The study appeared to show that unconscious brain activity associated with the action occurred half a second before the participants became aware of the decision to move, suggesting that decisions are first made by the brain and we only become conscious of them after a delay.

However, Libet's study had several flaws.[5] One was that it relied on the volunteers' own recording of when they felt the intention to move. However, this means a delay may have occurred between the impulse to act and the recording of it. It also was not clear that the EEG activity even represented the decision to move. Instead, it might have been linked to the act of paying attention to the wrist or a button, or just the expectation of movement. Indeed, in a later study in which volunteers were asked to press one of two buttons in response to computer screen images, their EEG activity changed even before the images came up on the screen, suggesting that it was not related to deciding which button to press.

If human free will cannot be so easily dismissed and yet we also recognise that a substantial part of consciousness involves unconscious biological mechanisms in the brain, how might we reconcile these two things? Can we also find evidence that free will is also present in, say, non-human primates or is it a specifically human characteristic?

Conceptual framework

I have said previously that a key difference between humans and other species is that only we have a conceptual framework, based on language, that allows us to reflect and reason, not only about our own situation but also that of other human beings, and also to come up with abstract ideas such as the meaning of life and what else lies out there in the universe. Yet although non-human primates lack such a conceptual framework for consciousness, based on language, they do have relatively developed brains compared to other animal groups, and also complex and sophisticated forms of society, so how might this affect their consciousness and does it mean that they might also possess a kind of free will?

In addressing this question, it is worth considering the studies that I have previously mentioned, by Earl Miller and his colleagues, that investigated working memory in monkeys. One potential criticism of the studies is that the monkeys were trained to respond in the ways they did. Although they made one choice over another in their responses, this was because they had learned they would receive a reward for doing so. Yet against this criticism, it is clear that the monkeys were making sophisticated selections. So not only were they able to remember images, and also sequences of images, they could even select images into categories based on abstract properties of such images. In fact, such abilities are perhaps not so surprising if we consider the challenges faced by a primate in the wild.

Studies have shown that primates in the wild live in large and highly complex societies, and individual primates must navigate a complicated and shifting network of alliances and rivalries to gain the best access to food, shelter and sexual partners.[6] We have also seen that non-human primates share with humans the ability to distinguish faces. They also seem to be skilled in recognising the nuances of social behaviour, by facial expression, by the sounds made by other primates in their group and by other forms of body language.

All of this means that the ability to mentally juggle various pieces of incoming sensory data, not only about potential prey or predators and other challenging features of a natural habitat, but also what other primates in the group are doing, could have been selected by evolutionary mechanisms. This also helps to address another concern I have previously raised, which is how the prefrontal cortex can play an 'executive' role in brain function. For although we lack the precise details, presumably this brain region evolved to play a coordinating role in juggling incoming sensory data because doing so brings rewards.

In the working memory studies, such a reward was given to the monkey by the experimenters; in the wild, the reward would be access to food or a sexual partner, but maybe also the emotional effects of being acknowledged and valued by other primates in the group. What this means in terms of brain mechanisms largely remains to be discovered, but it is likely that the prefrontal cortex will also be receiving feedback from reward centres in the brain and those dealing with emotions. If so, this suggests that far from the prefrontal cortex being only 'top-down' in its interactions with the rest of the brain, signals will be coming into this brain region from many other areas besides the sensory cortex that will influence its activity. It seems likely that these will also be coordinated by brain waves.

Primate choices

If this is how a non-human primate's brain allows it to make one type of potentially quite sophisticated choice over another, can we equate this with free will? On the one hand, we might agree that a monkey or ape is exercising its free will in making a specific choice that is related to the challenges and opportunities within ape society. On the other hand, however, we need to consider how much greater a hold instinctual behaviour has in non-human primates. In fact, there is some evidence that chimps can supress their natural urge to cry out in the presence of something that looks tasty to eat, in certain circumstances.[7] I am thinking here of an observation made by the zoologist Jane Goodall about two chimps she was studying, one, named Beethoven, dominant to the other, Dilly. Beethoven exerted his dominance by eating any bananas that were available, so one day Goodall secretively placed a banana out of sight of Beethoven, but within the view of Dilly. Intriguingly, according to Goodall, Dilly refrained from making a movement towards the food, or showing any sign of seeing it, until Beethoven had gone to another part of the chimp's enclosure. Studies by the anthropologist Brian Hare of similar chimp interactions over food in a more rigorous experimental setting also confirm a certain amount of control by chimps over their actions.

If this is the situation in non-human primates, how does this relate to human volition and free will? To address this question properly, we need to return to language. Previously, I have stated that human consciousness is qualitatively different from that of other species, even our primate cousins, because only our consciousness has been transformed by language, and as a consequence only human beings are able to think conceptually. In fact, the sophisticated nature of the categorisation that seems to have evolved in primates compared to other mammals may have provided an important, pre-verbal boost to the development of language-based conceptualisation in our ancestors and allowed them to make this leap. However, it was a leap nevertheless and suggests that human life choices will be qualitatively different, and more explicit than in, say, a chimpanzee, because only human beings are able to rationalise such choices and make value judgements based on our ability to think conceptually. At the same time, because of the social dimensions of our human minds, our choices in life and the value judgments that we make are influenced by the specific society that we live in, in ways that I will look at in more detail later in this book. The human ability to conceptualise is also relevant to the second topic I wanted to discuss in this chapter, which is whether the feeling we have as individual human beings of being a unique self is real or an illusion and also whether this feeling is specific to human beings.

Not everyone agrees that only humans have a concept of self. One approach that has challenged this idea involves studying chimpanzees' responses to their reflections in a mirror. In the first study of this type, by the psychologist Gordon Gallup in 1970, chimps that had never been exposed to a mirror were shown one and their behaviour observed. Initially, the animals reacted by making threatening noises and gestures; in other words, they seemed to believe that the mirror's reflection represented another chimp. However, after several days, the chimps began to try and touch parts of their body that were invisible in the mirror and even move food towards their mouths while watching the reflection. In later studies, chimps trained to respond in this way had their foreheads marked with an odourless dye and when the animal was placed in front of a mirror they were more likely to touch the dye spot. Such findings are said to demonstrate that chimps have self-conscious awareness.

Do these studies really show that chimps have a concept of self or can we find other explanations for the findings? One potential flaw in such studies is that while they show that chimps seem to associate parts of their bodies, and movements they make with their bodies, with themselves, this does not necessarily mean they have a concept of self. Instead, it could be rather that they have learned to associate the movements made by a reflection in a mirror with actions that they have performed. A second potential flaw in the studies is that the chimps were all extensively trained and rewarded for 'correct' choices, while animals that did not 'perform well' were removed from the study. This could mean that the experimenters selected for outcomes that appear to indicate self-awareness, but really just show that animals can make complex associations, given enough time and a reward, while animals that did not behave in line with the experimenters' expectations were excluded.

Language restructuring

With such potential flaws in studies that have purported to show self-conscious awareness in apes, and given that humans, due to our capacity for language, are the only species that can reveal what it feels like to have such awareness and everything that goes with that, it seems a good time to move to thinking specifically about human consciousness. In particular, I want to consider how what I have said about the importance of language and other cultural 'tools' in shaping human consciousness relates to the kind of novel mechanisms that I have mentioned in discussing recent studies of brain function.

In this respect, an important point to make is that evolution tends to build on existing structures and there is no reason to believe that the human brain is any different in this respect. It is therefore likely that brain waves of different frequencies play the same coordinating role in the human prefrontal cortex and other interacting brain regions. However, what will be different is that language will add an additional element of coordination, which is transformative in terms of its effect on our human thought processes.

I have already mentioned how inner speech, and more generally inner symbols of different kinds, provide a conceptual framework for structuring our thoughts. Somehow this must therefore have a basis in brain mechanisms. Given that the prefrontal cortex is already playing an executive role in non-human primates, presumably this means that inner speech adds a further layer of coordination on top of this. How this works at the level of brain mechanisms is still unclear, but I would imagine that inner speech will have an input into brain function via the same mechanisms involving brain waves of different frequencies, as I have already described. It is also likely to be the case that such language-based coordination will somehow interact with more ancient evolutionary mechanisms, such as the capacity for categorisation based on non-language-based mechanisms that exist in non-human primates, but also even older mechanisms that involve brain regions involved in emotional responses.

Something I do find exciting about recent discoveries about the role of brain waves as coordinators of different brain regions, and as 'executors' of brain function, is that they may help us address some centuries-old debates about the material nature of human consciousness, and how this might differ from that of animals, in a novel way. Take, for instance, one of the first attempts to provide a material view of consciousness, the 'blank slate' view of the mind put forward by John Locke and David Hume in the 18th century. This viewpoint saw each individual human consciousness as just the accumulation of experiences acquired since birth. Yet, a problem with this view was that it did not explain how each individual mind feels like a unified, distinct phenomenon, rather than just a mass of unconnected experiences. However, we now have a potential solution to this 'binding problem'.

If brain waves of different frequencies segregate within the brain sensory signals relating to, say, the shape, texture, colour and other characteristics of objects in the outside world, but also combine them together into a unified whole, then this explains how we bind together these different aspects of experience into such a unified whole. If we then factor in the effect of language and other cultural tools, and how they have transformed human consciousness, we can see how this provides another level of binding, through the conceptual framework that this provides. This surely means that our sense of self is not an illusion, but rather a very real phenomenon based on the binding role of brain waves and the extra element of unity based on conceptual thought. In other words, our sense of self is based both on brain biology and also on the concepts of self versus other. Our sense of self and also our feeling that we have free will are based on the fact that, as human beings, we are active agents of our destiny and also make rational choices. Having said all this, we also need to ask why human beings are prone to behaviour that often seems decidedly irrational. To answer this question, we need to look more closely at the unconscious mind.

CONSCIOUSNESS AND THE UNCONSCIOUS 16

The concept of the unconscious mind has a chequered and controversial history in psychology. Of course, in some respects the idea of an unconscious mind is not controversial at all. Control by the brain of our breathing, our heartrate and a vast range of other bodily functions that keep us alive and well all occur in an unconscious fashion each moment of our lives. We have also seen how procedural memory – that form of memory that allows us to learn new motor skills, like riding a bike or playing the piano and of which the cerebellum is a particularly important mediator – largely involves unconscious actions.

There is, however, another more controversial view that sees even our conscious thoughts as being heavily influenced by unconscious feelings and desires, in ways that undermine the idea that as individuals we are fully in conscious control of our thoughts and actions. Such a viewpoint became particularly popular in the Romantic movement of the late 18th and early 19th centuries.[1] The Romantic writers William Wordsworth, Samuel Taylor Coleridge and Thomas De Quincey saw unconscious feelings as an important source of human creativity and believed that such feelings might be given freer rein in dreams or under the influence of mind-altering drugs. A famous example of this is Coleridge's claim that he wrote his poem 'Kubla Khan' after waking from an opium-influenced dream. Meanwhile, the Romantic philosopher Carl Gustav Carus saw 'the key to an understanding of the nature of the conscious life of the soul ... in the sphere of the unconscious'.[2]

Despite such important antecedents, undoubtedly, the individual that most people today would associate with the idea of an unconscious mind is the psychologist Sigmund Freud. He viewed the human mind as consisting of three components – the id, the ego and the superego – and he believed that the id forms our unconscious drives and seeks only to satisfy pleasure; the ego is our conscious perceptions, memories and thoughts that enables us to deal effectively with reality; and the superego tries to control the demands of the id through socially acceptable behaviours.[3] Based on this standpoint, Freud argued that repression of unconscious sexual desires is a major source of mental disorders.

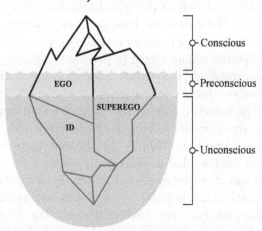

Figure 12. Freud's model of the mind (the iceberg methaphor).

Today, opinions about Freud are highly polarised. For while a recent BBC documentary hailed him as a 'genius of the modern world',[4] the biologist Peter Medawar saw psychoanalytic theory as 'the most stupendous intellectual confidence trick of the twentieth century'.[5] The thrust of his argument was that Freud was an opportunistic self-dramatiser who deliberately misrepresented the scientific basis of his theories.

Dynamic unconscious

Personally, I share the scepticism about the scientific validity of many of Freud's ideas, for instance, his view of the mind as a hydraulic system in which different thoughts and desires literally raise or lower pressure within the mind and his now outdated and often reactionary views about the basis of women's sexual feelings and homosexuality. Yet I believe that Freud was correct in his recognition of the importance of the unconscious and insightful in some of what he said about the dynamics of unconscious thoughts. An important question to ask, therefore, is how we might explain the material basis of a repressed unconscious in terms of the view of consciousness I have been developing in this book. The answer, I believe, involves going back to the importance of inner speech, or more generally, inner symbols, as a critical aspect of the human thought process.

I have talked about the potential differences between inner and external speech, our understanding of which is of course limited by the difficulties in directly studying this phenomenon that occurs inside our heads. So inner speech appears to be much more rapid than external speech, with the psychologist Charles Fernyhough estimating that it may be as much as ten times as fast as the external variety. One reason for this is that internal speech does not require the rate-limiting apparatus of the mouth, tongue, throat and vocal cords to be voiced. Another is that inner speech seems much more telegraphic than external speech. This is because, although I have argued that inner speech has the characteristics of a dialogue, it is nevertheless an internal one, and so much can be assumed by the two sides of the dialogue without being explicitly voiced. Linked to this, inner speech seems more fragmentary and less fully formed than external speech.

Inner speech is in general more fluid in meaning than external speech.[6] This was something noted by the psychologist Lev Vygotsky when he stated that 'the senses of different words flow into one another ... so that the earlier ones are contained in and modify the later ones'.[7] In fact, this is very similar to what Freud observed about dreams, namely that 'a single word, if it is especially suitable on account of its many connections, takes over the representation of a whole train of thought'.[8] This fluidity, which Vygotsky saw as a tendency in inner speech towards a 'preponderance of sense of a word over its meaning' has major consequences for our understanding of what we mean by the unconscious mind.

Far from representing some kind of separate entity in human consciousness as Freud viewed it, we can see the unconscious as more a manifestation of the fact that the inner dialogue in our heads is far from straightforward and instead reflects contradictions that can be difficult to express, never mind resolve, and also things that remain unsaid, even in our innermost thoughts, because of the impact this might have on our sense of right and wrong. In this sense, I think Freud was right to point to the clash in our minds between our innermost thoughts, feelings and desires, and what is seen as being socially acceptable. He was also correct in noting that one place where our inner thoughts are most fluid, and least held back by social norms, is in our dreams, which is why these often contain odd associations of individuals and scenarios we might hesitate to recite to others out loud.

The philosopher Valentin Voloshinov, whose insights into the nature of inner speech I have already mentioned, argued that the unconscious represents the least articulate part of the continuous dialogue taking place in our heads, because the themes with which it is concerned are those where the 'official' ideology of society is most at variance with a person's actual experience.[9] Viewed like this, it is perhaps no surprise that Freud became so focused on repressed sexuality as a central feature of the unconscious, since it is hard to think of an area of people's lives where the gap between public pronouncements and private feelings and practice are more divergent. This was particularly true in the late 19th and early 20th centuries when Freud was developing his ideas about the unconscious.

Contradictory consciousness

However, if we want to better understand the role of this aspect of the unconscious in human consciousness, we need to move beyond Freud's primary focus on sex and think more generally about all the areas where official and unofficial ideologies might clash within our heads. I also want to introduce another important concept, that of 'contradictory consciousness'. Particularly associated with the philosopher and political theorist Antonio Gramsci, this expresses the idea that people can hold two opposing opinions about a subject.[10] If we conceive consciousness as simply rational thought, then it is hard to see how a person could hold two mutually antagonistic views at the same time, but if viewed from the perspective of an unconscious mind, there seem to be ample possibilities for contradictory views to coexist in our heads, as they may be expressed so subliminally or in a disguised form that we may not be aware of the contradiction, or somehow gloss over it.

Ultimately, the unconscious aspects of human consciousness, and also its contradictory character, reflect contradictions in the world around us and our relationship with that world. With that in mind, let us now explore those contradictions, which means taking a closer look at human society, both in its original form and society today.

For the vast majority of our time on Earth, human beings have lived in small groups, hunting wild animals and gathering plants, grubs and so on. Therefore, getting a better idea of what it means to live in a hunter-gatherer society should give us insights into the sort of society in which human consciousness first evolved. To complicate matters, however, existing hunter-gatherer groups have been heavily affected by contact with the civilisations that dominate the world today. The archaeological evidence of prehistoric hunter-gatherer societies is also very sparse. In particular, while we might gain important insights into the type of tools our ancestors used, we have no way of knowing what was going on inside their heads.

What we have learned, however, is that hunter-gatherer society tends to be above all about cooperation, minimising conflict and ensuring that everyone's role is respected. One measure for ensuring this in the Ju/'hoansi 'bush people' of the Kalahari, is what is known as 'insulting the kill'. A Ju/'hoan man explained to the anthropologist Richard Lee, who carried out some pioneering studies of this group of hunter-gatherers in the 1960s, what this means: 'When a young man kills much meat, he comes to think of us as his servants or inferiors. We can't accept this ... so we always think of his meat as worthless. This way, we cool his heart and make him gentle.'[11] Such a measure is typical of a type of society that has no formalised leadership institutions and in which men and women enjoy equal decision-making powers, children play largely non-competitive games and the elderly, while treated with affection and respect for their skills and experience, enjoy no special privileges.

Such features of hunter-gatherer groups reflect the fact that such groups have minimal possessions and their existence is dependent on the animals they can kill and the plants and grubs that they can collect from the wild, so it would make no sense for one individual, or set of individuals, to be elevated to a higher status; indeed, this could be catastrophic for the group. Although we have less information about interactions between different hunter-gatherer groups, because of the scarcity of such remaining groups on the planet, the emphasis seems to be on similarly reducing conflict.

Given that egalitarianism, cooperation and reduction of conflict are key features of hunter-gatherer societies, and given that humans and our proto-human ancestors have lived in such a type of society for the vast majority of our time on Earth, how might this have affected our biologically evolved human consciousness? At the very least, it goes against claims that human beings are naturally drawn to inequality, competition, conflict and war.

It also means that the way that women are still perceived as inferior to men in many societies, and the very real gender-based imbalances in status and power that continue to exist in every country in the world, are not natural features of humanity, but a relatively recent development. The same is true of the idea that some people are of lesser worth because of the colour of their skin or some other physical characteristic. This means that if we want to understand the material basis of sexism or racism, we need to look beyond biology and at what has changed since those hunter-gatherer days.

Civilised states

In particular, we need to think about the effect that the move away from a hunter-gatherer lifestyle towards our current way of life centred on civilisation has had on human consciousness. In this respect, the decisive shift occurred about 12,000 years ago, when the Natufians, a group of hunter-gatherers in an area of the Middle East that is now the Lebanon, first made the move from hunting and gathering to agriculture.[12] Their actions led to a revolution that would transform humanity's interactions with the natural world, but also our relations with each other. For the agricultural revolution would in turn lead to the rise of the first city states and the birth of a human civilisation whose development continues to spread across the world, in ever greater complexity and technological power.

What have been the effects of civilisation on human consciousness? To answer this question properly we need to consider both positive and negative sides of the rise of civilisation. Take, for instance, the ancient societies of Egypt and the Roman Empire. I recently visited the pyramids near Cairo for the first time and although they are such an iconic and familiar sight, it was still quite breath-taking to experience them as real objects, not as images on a postcard or on some computer screen. Similarly, on a recent trip to Rome to visit a colleague, my host took me on a night-time tour of the city and walking up with him to the top of Capitol Hill to see the floodlit ruins of the Roman Forum brought home to me both the power and prestige, but also the beauty, of that ancient empire.

Yet one thing that all these ancient states had in common is that they were based on slavery. This was not some aberration of these societies but fundamental to the way they arose, and consolidated their power. For the first city-states were built on extracting value from the surrounding countryside by taxation and force, amassing increasingly larger tracts of land and possessions by warfare, and from such wars capturing human beings and turning them into slaves, who then carried out the bulk of work in these societies.[13]

How did all this affect human consciousness? In answering this question, one of the problems we face is the difficulty of imagining a totally different way of looking at the world than the one we have today. Take, for instance, the philosopher Aristotle. Today he is revered as one of the greatest philosophers of all time, and indeed I mentioned his insights into the nature of consciousness at the start of this book. Yet when discussing some of his fellow human beings, Aristotle stated that 'the use made of slaves and of tame animals is not very different; for both with their bodies minister to the needs of life'.[14] In other words, he seems to have seen slaves as little better than animals, objects to be owned and used.

The fact that so insightful and gifted a thinker as Aristotle could hold such an appalling viewpoint by today's standards, raises important issues about human consciousness. First, it shows how the structure of a particular human society can influence some quite fundamental human ideas and beliefs. After all, less than 10,000 years before Aristotle, humans only existed in hunter-gatherer groups organised along principles of strict equality. In that time period, which is tiny in evolutionary terms, it seems unlikely that anything substantial had happened biologically to the human brain. What clearly had changed was that a subsection of people – a class – had risen to the top of society and had not only developed an ideology that justified to themselves behaviour that today we would find both reprehensible and indefensible, but also, at the same time, created a ruling ideology that presumably most people in that society believed.

A second important point to be made in considering Aristotle and his defence of slavery is the contradictory nature of human consciousness. At least, that is the only conclusion I can draw from the fact that in his statement Aristotle directly equated slaves with animals, yet surely in his personal dealings with his own slaves, he must have recognised that here were individuals who could talk, respond to detailed spoken commands and carry out complicated tasks, in ways that would be impossible for an animal. Added to this, in ancient Greek and Roman society, on occasion, a slave could gain so much responsibility, respect and even wealth, that they were able to gain their freedom.[15] In this circumstance, surely there would be some contradiction in dealing with the fact that an individual previously classified as equal to an animal was now considered human. This raises the question of whether individuals like Aristotle were conscious of such a contradiction, or somehow managed to gloss over it, in their conscious thoughts.

Today we live in a very different type of society in which forms of slavery still exist in some pockets of the globe, but are generally not seen as something defensible, and definitely not a normal state of affairs.[16] Modern society is also very different from feudal society, the type of society that followed the ancient slave societies. While slavery no longer existed in feudal society, the subjugation of the peasant by the aristocracy was both justified by the ideology of the time, in which religion had an important role to play, but also enforced by military might. In contrast, the revolutions that occurred in the 17th and 18th centuries in the Netherlands, England, the USA and France that first brought the capitalist class to power, were inspired by statements such as 'all men are created equal [and have] certain unalienable Rights ... among these are Life, Liberty and the pursuit of Happiness', or the more succinct 'Liberté, égalité, fraternité'. Yet a glance at the world today reveals striking inequality, illustrated by a recent finding by the charity Oxfam that the world's richest 26 individuals own as much as the poorest 50 per cent of people on the planet.[17] So how do these contradictions in modern society affect human minds and human behaviour? For that, we need to consider more deeply the dynamics of modern, capitalist society.

MODERNITY AND ITS CONTRADICTIONS 17

At the heart of modern society lies a contradiction. On the one hand, the capitalist system that now dominates the world is not just the most dynamic in world history, but in its achievements dwarfs anything realised by past societies. So while we can admire the Egyptian pyramids, the Parthenon in Athens, the Colosseum in Rome, the Great Wall of China and the Gothic cathedrals of Europe, in just a few hundred years, capitalism has created towering skyscrapers, bullet trains, supersonic planes and technologies that have transformed the lives of ordinary people in ways that would have been unbelievable to people only a century ago. There are also antibiotics, keyhole surgery, gene editing, bone marrow transplants, laptop computers, smart phones, the internet ... the list goes on, and our technologies now not only exist on Earth but beam back images from Mars or Pluto.

Yet while modern society is highly dynamic, it also has a tendency to instability, crisis and ultimately, unsustainability. Not that the instability aspect of capitalism is seen as a bad thing by some commentators. For them the whole point of new technologies, such as social media or gene editing, is to destabilise existing ways of working, because they believe that sometimes society needs a shake-up if it is to move forward. There are even those who believe that stock market crashes, depressions and slumps have their role to play by eliminating moribund companies and institutions and favouring dynamic ones, in a kind of 'survival of the fittest' view of economics. Yet I hope it would be hard to find anyone willing to say that runaway global warming is a good thing, or the prediction that if current rates of extinction due to human activities continue, half of Earth's species will be extinct by 2100.

If these are some features of modern society, how might it affect human consciousness? On the face of it, current society in a typical liberal democracy is very different from the ones our ancestors lived in 2,000, or even 200, years ago. In such a democracy, human beings are no longer bought and sold as slaves and you would not be executed for making fun of a nobleman. Both sexes have the right to vote and increasingly women can be found in the most highly paid and powerful positions in society, although overall there is still a significant gender gap in both pay and social influence. In the past, only a tiny proportion of people went to school; today, in a country like Britain or the USA it is obligatory, and substantial numbers also go to university. In theory, anyone, male or female, black or white, can become British Prime Minister or US President, as shown by the elections of Margaret Thatcher and Barack Obama.

Profit system

Yet not only are liberal democracies not the majority state of affairs – with a recent study claiming that around 70 per cent of people in the world live in some form of dictatorship[1] – but even in such democracies there are growing levels of inequality. Added to this are reports of how stressed and dissatisfied many people feel about their jobs globally. For instance, a recent study by Microsoft surveyed 20,000 people in eleven countries between July and August 2022, and reported that 50 per cent of employees and 53 per cent of managers feel 'burned out' at work.[2] Add to that the current 'cost of living' crisis, based on escalating food and fuel prices, with many low-paid workers saying that they are facing a choice between heating their homes or feeding their families, and we have what looks like a serious problem both in terms of people's living standards and their mental health. So what is the material basis of these problems and how might it affect people's minds?

In seeking to understand the underlying dynamic of modern capitalist society, it is worth looking at the insights of some of the earliest commentators on this type of society. Today, the political economist Adam Smith is widely recognised for identifying three key features of capitalism that distinguished it from previous economic systems.[3] First, under capitalism, anything can become a commodity if someone is willing to buy it. Second, this is a system of two opposing forces: the producer seeks to maximise their profit, while the buyer wants to pay the lowest price possible. This leads to competition between producers, either to lower their prices but get less profit, or to innovate their production process to make it more profitable. Third, a key way in which production can become more efficient is by dividing up a highly complex production process, requiring multiple skills for a single individual to perform, into a number of separate operations, each requiring much less skill. This can boost profitability as well as efficiency as less skilled workers can be paid less.

The industrialist Henry Ford perfected such a division of labour in his car factories by developing an assembly line system, which introduced another aspect to the process: through speeding up the assembly line it was possible to pressurise each worker to perform their part of the manufacturing process at the maximum speed possible.[4] A sign of how this works in modern times can be seen at one of the delivery warehouses of the tech giant Amazon; in these, human 'pickers' move around the warehouse on a predetermined route, collecting items for delivery and scanning each with a handheld scanner. Not only does this record when an item has been collected, it also allows managers to monitor the pickers' work speed and to put pressure on them to increase this if necessary.[5]

While division of labour and assembly lines can boost both efficiency and profitability, they come at a human cost. Because workers in an assembly line only perform one small task in the overall production process, and because they are continually under pressure to work faster, far from feeling fulfilled in their work, this can come to seem like a heavy burden. Adam Smith recognised this when he stated that such a form of production required the worker to give up 'his tranquility, his freedom, and his happiness'.[6] This view was echoed by another economist, Karl Marx, when he stated that 'the worker feels himself only when he is not working, when he is working, he does not feel himself … his labour is therefore not voluntary but forced … its alien character is clearly demonstrated by the fact that as soon as no physical or other compulsion exists, it is shunned like the plague'.[7]

Alienated society

However, Marx went further than Smith in seeing such 'alienation' of a worker from their labour as part of a much more pervasive tendency within modern society, that affects not only people's attitude to work but also their relations with each other, and even their view of themselves.[8] So how does alienation affect people in these different kinds of ways and what is the relevance of all this to our quest to understand human consciousness?

We have already seen how one way that modern production methods can affect people's sense of themselves is through the atomisation of work. In contrast to the pride of a skilled craftsperson in their creations, a worker on an assembly line may feel that they are merely a small cog in a large machine and not in the positive sense of contributing to a greater goal, but the negative one of being treated as if they were just a piece of machinery.

This feeling can also affect those at the top of society, for it can lead to those who make decisions about the running and financing of companies to view workers as expendable as an obsolete piece of machinery or little more than items on a balance sheet. Use of apparently neutral modern business terms like 'down-sizing' or 'rationalisation' play a part in masking the human consequences of, for example, sacking workers or closing a factory.

Such dehumanisation of workers is also linked to two other tendencies of modern society, namely that everything can become a commodity under capitalism and the fact that competition is central to this form of production. In terms of work, these tendencies reinforce the idea that workers are there to be bought and sold, but it can also lead to workers seeing each other as the primary obstacle to their success at work or even having a job. This can particularly become an issue if, for example, immigrant workers from another country are thought to be undermining wages and conditions by accepting less than the norm. It does not take much for such feelings to be linked to prejudices, which may be stoked by sections of the establishment or the media, about a person's skin colour or religion.

Meanwhile, competition is such a key part of the way capitalism works that, despite the fact that cooperation is a central feature of the original hunter-gatherer society in which human beings have lived for the vast majority of our time on Earth, from school onwards we are told that the only way to find success in life is through competition, whether passing exams while at school or university, or showing competitive qualities in the workplace.

Yet in direct contradiction of this idea, far from social mobility increasing in recent decades, the opposite has been the case, and in general wages have stagnated and jobs have become more precarious, which might lead workers to question the ideology they have been brought up to believe, but could also make some blame themselves for their situation in life.

Alienation affects more than people's attitude to their work and other workers; it also has a crucial effect on all kinds of other aspects of their lives and society in general. Take, for instance, the fact that anything can potentially become a commodity under capitalism. This can affect people's attitude to sex, to the extent that other human beings may be viewed as objects that can be bought or sold. Of course, one could argue that there is nothing new about prostitution – after all, it is often called the 'oldest profession'. Yet where capitalism excels is in scale and reach compared to past economic systems and also in its tendency to continually transform different areas of society through the introduction of new technologies, and the sex industry is no different in this respect. In fact, this process has been particularly accelerated with the rise of the internet and social media. So not only can prostitutes make physical contact with clients through a variety of websites and social media platforms, but increasingly phone sex and video sex can also be bought via the internet. And if someone looking for sexual gratification prefers to remain a voyeur, the internet now hosts a bewildering variety of pornographic videos, photos and cartoons.

The rise of the internet sex industry may also be having a more pervasive influence in overall perceptions of what it means to interact sexually with another human being. It can normalise the idea that finding a date who may be a sexual partner is just a matter of a rightwards swipe on a phone screen, that abusive sexual behaviour first viewed on a porn site is therefore permissible with a partner or the general idea that other human beings are objects whose main purpose is to be viewed voyeuristically or used and then discarded.

Internet influence

In fact, the internet is beginning to have a far more general influence on human society for increasingly, through sites such as Facebook, Twitter, Instagram and TikTok, a person's travels, visits to restaurants, unusual lifestyle or peculiar quirks, or indeed any aspect of their lives, can be monetarised via the number of likes or subscribers that an individual accumulates, because of the opportunities this provides for advertising.

All of this means that commodification has extended into areas of human society and an individual's life with a reach and on a scale that an entrepreneur of, say, the early 20th century could have only dreamed of, but it is also part of an increasingly common culture in which the success of any individual is judged both by their social media presence, as well as how aware they are of new developments on the internet, and respond to them.

While this may be a particular issue for celebrities and 'influencers' who primarily make their living through the internet, it is increasingly the case that it is a sign of normality to be connected to the internet throughout a person's waking hours, both in terms of getting the latest updates from all that is out there of interest and displaying, to all those interested to view and listen, what that person has been up to for the past 24 hours.

The modern pressure to be always connected via the internet is an issue for all human beings, but particularly so for children and young adults. This is because their brains are still developing, so they are more open to novel ways of doing things. Given that during adolescence sexuality is blossoming and friendships can be quite fluid, opening up new possibilities, but also a strong desire to be liked and respected by peers and not 'left out' of activities, this can put pressure on young people to maintain an active presence on social media. However, due to the competitive ethos in capitalism, this also exposes young people to inflated images of what a normal body should look like – sometimes literally so in the case of breast, buttock or lip enlargements – and this can have a negative and destabilising effect on young people's feelings about their own body, particularly in young women.

If these are some of the pressures and contradictions of modern society, how does this relate to the dynamics of human consciousness and brain mechanisms already mentioned?

One feature of the human brain that may explain why people can come to accept and even acquiesce in an unhappy, exploitative situation at work, and even see other workers as the primary threat rather than as potential allies in campaigns for better pay, working conditions and job security, or buy into the idea that competition is the natural state of affairs for human society, is the way our brains have evolved to respond to living in a hostile natural environment. I have mentioned that life on Earth is based on two principles: survival and reproduction. To do both effectively, our brains have evolved to work in two often opposing directions. One is a tendency to keep doing the same things that have kept a species alive and reproductive across many generations; the other is the ability to respond rapidly to any threat to that status quo. The first tendency, which encourages sticking with a certain pattern of behaviour and is rooted in the more primeval parts of the brain, may explain why it can take a considerable change in a person's job satisfaction or happiness as an individual to make them think about challenging the status quo.

As for the second tendency, two brain regions that I have already mentioned, the prefrontal cortex and the amygdala, are relevant. As a primary mediator of emotional responses, the amygdala is one of the main ways that our brains respond to potentially threatening situations, and there is increasing evidence that many unconscious fears about a perceived threat of people of a different colour skin or sexual persuasion, or even male fears about female equality, involve the amygdala. Therefore, although inflammatory speeches by politicians or media articles may enhance such fears, they are helped by deep-seated impulses in the human brain that may be only partially operating at a conscious level.

Despite this role of the amygdala in propagating unconscious fears and prejudices, an important aspect of being human is that not only did our ape ancestors' brains evolve sophisticated 'higher' brain regions like the prefrontal cortex, but language has transformed our brains in ways that I have already outlined, and this can play an important mediating role in allowing us to rationalise our situation in life and potentially change it. Later I will be looking at how this can particularly occur at times of great social change. However, before I do that, I want to consider what happens when the stresses and pressures of life become so great that a person becomes unable to cope, and to look more generally at mental distress, its biological and social roots and why so-called mental 'disorders' have become such a problem for individual humans and human society as a whole.

SANITY AND MADNESS

18

For a species with such a unique gift of conscious mental awareness, it is remarkable how often that gift can turn into a burden. Mental distress is recognised as a major problem in modern society, with around one in four people treated for a psychiatric disorder in Britain, and a similar pattern in North America and mainland Europe.[1] A central difficulty in diagnosing and treating mental disorders is reaching a level of agreement about their underlying basis. One view is that mental disorders are primarily biological in origin and it is just a case of identifying the genes responsible. Yet despite resources like the Human Genome Project and studies that have 'mapped' the genomes of people with disorders like schizophrenia or clinical depression, the problem has been that far from identifying a few clear associations with specific genes, such findings have challenged all those who imagined the link between mental disorders and the human genome would be a straightforward one.

For instance, a recent study of the genetic basis of schizophrenia in 36,000 patients found association with 108 genomic regions.[2] The discovery of so many genomic links confounds previous expectations by some people that there would be just a few 'schizophrenia genes'. It also raises questions as to whether schizophrenia is really a single disorder if so many genomic regions are involved in its development.

Other studies have sought to identify biological differences in the brains of schizophrenics. For instance, the findings of one recent study suggests that schizophrenia may be trigged by an inappropriate immune response involving a surge in the number and activity of glial cells in the brain.[3] One role of glial cells is to help fight infection, so a surge in their numbers often occurs in response to an infective agent in the brain, but they are also involved in 'pruning' unwanted connections between neurons. However, in schizophrenia, the pruning may be overly aggressive, leading to vital neuronal connections being lost. In line with the idea that a brain mechanism that evolved to protect the brain has instead become overly aggressive and triggered this mental disorder, other recent studies have suggested that schizophrenia seems often to be associated with inflammation in the brain.

So what might trigger such changes in the brain? Some recent studies suggest that genetic differences in an individual make them particularly susceptible to a breakdown in normal brain function. However, there is also increasing evidence that social trauma and abuse can also have quite profound effects on brain biology and even on the way that genes function within the brain. It is therefore worth looking at the type of traumatic social events that may be a precipitating factor in causing the biological changes that may lead to schizophrenia.

Communication problems

In addressing this question, it is useful to look at a very different way of explaining schizophrenia than one that focuses primarily on genes and brain biology. This view of schizophrenia is particularly associated with the psychiatrist R.D.

Laing. He believed that a key factor in the development of this mental disorder is major problems of communication within the schizophrenic's family. Typically, the person who becomes schizophrenic is caught in a situation in which messages from other family members are deeply contradictory; however, the contradiction is never brought into the open and the person is unable to leave the field of interaction with such family members.[4] More generally, studies suggest that sexual, physical and emotional abuse, and a variety of other types of adverse childhood experiences may be important factors in the development of a schizophrenic.

If schizophrenia may be due both to differences in individual biology and a dysfunctional or harmful social environment, how does it actually manifest itself in the brain? And how is this related to the view I have been developing in this book, which sees our minds as mediated by language and culture? Following such a viewpoint, since the development of human consciousness follows a specific pattern in its path of 'mediation' by society, so its descent into a disordered state like that in schizophrenia might also follow such a pattern, but in reverse. In line with this, studies by the psychologists Eugenia Hanfmann and Jakob Kasanin have revealed that there seems to be a breakdown in conceptualisation in schizophrenics.[5] While such individuals may retain an ability to communicate with the same words as an unaffected person in terms of the objects they refer to, their words will exist in totally different systems of meanings.

If different biological origins for schizophrenia might manifest themselves in common regressive behaviour at the conceptual level, can this help us better understand the social triggers that may help to induce a schizophrenic state? In this respect, it is interesting that a study by the psychologist Bjorn Rund found that non-paranoid schizophrenics, but notably not paranoid schizophrenics, generally came from families that displayed abnormalities in their patterns of communication.[6] For instance, the parents of such schizo-phrenics were often highly egocentric in their behaviour and tended to ignore what their children were saying, interrupt them or send mixed messages.

Rund has particularly focused on attention, the process whereby an individual's awareness becomes focused, like a spotlight, on a subset of what is going on in their head or their environment. Normally, voluntary forms of attention emerge as 'people who surround the child begin to use various stimuli and means to direct the child's attention and subordinate it to their control'. However, in a family situation in which normal communication has broken down, Rund believes that the child has 'no opportunity to establish a firm way of focusing attention on the relevant stimuli in a given situation, because it will never know which are the relevant stimuli. Instead, the attentional styles that are internalized will be characterized by a steady wandering from one stimulus to another, in a search for the most relevant one.' Rund also believes that the 'emotional climate' within a family could affect the development of schizophrenia. This is surely even more the case if physical and sexual abuse are a part of family life. Anxiety, insecurity and instability therefore tend to be a feature of the families in which schizophrenics grow up.

Rund has left open the possibility that if genetics play a role in the genesis of schizophrenia, this could be a factor in why the condition tends to run in families. For not only could a tendency towards schizophrenia be passed down directly, but even unaffected family members might still display abnormal behaviour linked to their genetics, and such behaviour could help trigger the condition in a vulnerable individual in the family. However, it is also likely that other factors outside the family – racism, sexism, homophobia, problems at school or work or simply the pressure of living in a modern capitalist society – may combine to precipitate the disorder in a specific individual within that family.

Disordered chemistry

A similarly complex interaction between 'nature and nurture' seems to be also the case for another common mental disorder – clinical depression. What seemed like a major step forward, both in our understanding of this disorder and how to treat it, occurred in the 1950s, when evidence emerged that depression might be due to low levels of the neurochemical serotonin. In particular, neuroscientists discovered that chemicals that alleviated the symptoms of depression seem to work by inhibiting the reuptake of serotonin into certain neurons in the brain. These chemicals, called selective serotonin reuptake inhibitors (SSRIs), have been a huge success for the pharmaceutical industry, with Prozac, one of the most well-known SSRIs, earning Eli Lilly, the company that developed it, billions of dollars.[7] Yet recently, doubts have surfaced about the effectiveness of SSRIs, the mechanisms by which they affect brain function and potential long-term side effects.

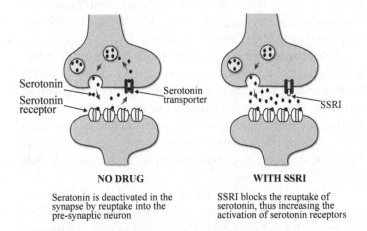

Serotonin

Serotonin receptor

Serotonin transporter

SSRI

NO DRUG

Seratonin is deactivated in the synapse by reuptake into the pre-synaptic neuron

WITH SSRI

SSRI blocks the reuptake of serotonin, thus increasing the activation of serotonin receptors

Figure 13. Proposed mechanism of action of SSRIs like Prozac.

One concern is that a recent study found that the ability of SSRIs to combat mild or moderate depression appeared no better than a placebo, although this did not generally appear to be the case if the treated individual is suffering from a more severe form of depression. Another worrying aspect of SSRIs is that while these drugs are only meant to be taken for a limited period, in practice, many patients remain on them for years, and for some it can be very difficult to stop taking them, such are the severe effects of withdrawal, which for one patient included 'severe shakes, suicidal thoughts, a feeling of too much caffeine in my brain, electric shocks, hallucinations, insane mood swings'.[8]

A major problem in determining the true effectiveness of SSRIs for the treatment of depression is how little is understood about the biological basis of this disorder and how drugs like SSRIs alleviate it. In particular, doubts have surfaced about whether depression is primarily caused by low levels of brain serotonin. The psychiatrist Alan Gelenberg thinks 'there's really no evidence that depression is a serotonin-deficiency syndrome. It's like saying that a headache is an aspirin-deficiency syndrome'.[9]

If doubts are being cast on a simplistic 'chemical imbalance' model of depression, are we getting closer to better understanding this disorder? One new theory of depression sees it as due to a lack of new neurons being regenerated, with lower serotonin levels being more a consequence than cause of depression, since when the brain stops making new neurons, or fewer neuronal connections are formed, this reduces serotonin release.[10]

Other recent studies have suggested a link between depression and inflammation.[11] I mentioned earlier how abnormalities in how glial cells respond to brain inflammation may be one factor in the development of schizophrenia. In depression, inflammation may affect the brain in a different way, sparking feelings of hopelessness, unhappiness and fatigue. Such feelings may be caused by the immune system failing to switch off after a trauma or illness, and triggering a severe version of the low mood people can experience when fighting a viral infection. Indeed, studies suggest that treating inflammation can alleviate depression, while treatments that boost our immunity to fight illness can be accompanied by a depressive mood; for instance, many people feel down after vaccination.

What about the possibility that certain individuals are biologically prone to depression? One problem investigating such a possibility is that depression is a common disorder, with more than 350 million people affected globally.[12] The disorder's symptoms and severity can vary widely from one person to the next and also between men and women. This means there may be a whole spectrum of different conditions that are being grouped together that have quite different biological causes, complicating genetic analysis.

To address this, a recent study focused on depressed women in China.[13] The researchers reasoned that because depression tends to be underdiagnosed in that country, women identified as clinically depressed might have a particularly severe form of the disorder. Indeed, those taking part in the study were at the most severe end of the spectrum. The study identified variations in two genes – LHPP and SIRT1 – as being clearly different in many of the depressed women. Although LHPP's function in the cell remains unclear, recent studies suggest that SIRT1 may protect neurons from ageing and death,[14] so this could be important given that, as we have seen, depression has been linked to inflammation and to defects in the brain's ability to regenerate itself through neurogenesis.

A complicating factor for any attempt to identify a biological basis to depression is that, just like the situation for schizophrenia, the social environment appears to play an important role in the genesis of depression. In 1978, George Brown and Tirril Harris studied the incidence of depression in London and concluded that the best predictor of this disorder is being a working-class woman with an unstable income and a child, living in a tower block.[15] One conclusion is that a stressful environment is far more likely to precipitate depression than a less stressful one. However, another factor that might trigger depression, given what I have said previously about the importance of social interaction in human society, is some disruption to such interaction. Indeed, Brown and Harris found that 89 per cent of the depressed women they interviewed had suffered a life-changing event such as a family bereavement, a marriage break-up or the loss of their job after a period of full employment.[16] While all of these events would presumably be highly stressful, they could also lead to a significant reduction in an affected individual's social interaction.

None of this means, however, that biological susceptibilities for depression can be ignored. Moreover, it would be a mistake to believe that the influence of biology and environment are easily separable, for, as we have seen, it is becoming clear that the environment can directly affect the genome through epigenetic mechanisms.

Ultimately, if our understanding of schizophrenia, clinical depression and also other mental disorders is to progress and hopefully in the process lead to new forms of diagnosis and treatment, this needs to be linked to a better understanding of how the brain works. Here it is relevant to consider how some of the new insights from neuroscience, which I have been discussing in this book, might relate to schizophrenia and other mental disorders.

Earlier I mentioned studies showing that brain waves of different frequencies play an important role in coordinating the activities of different brain regions. Such findings raise the interesting possibility that mental disorders may occur due to a breakdown of such coordination, but in a way that distinguishes one disorder from another. For instance, it may explain why multiple regions of the genome have been linked to a condition like schizophrenia, because although mutations in very different genomic regions may affect different parts of the brain, the end result could be a breakdown in overall brain function that manifests itself in the symptoms by which we tend to define a schizophrenic.

Such an interpretation also allows us to explain a well-known symptom of schizophrenia – the tendency to hear 'voices',[172] for a major claim of this book is that our innermost thoughts find expression through 'inner speech'. This viewpoint sees individual thought as a kind of internal dialogue, with the different voices within that dialogue being drawn from our social interactions, and individuality itself as a boundary phenomenon that is both a product of such social interactions and the biology of the individual's brain.

Disunited self

To some extent this viewpoint leads to a conundrum, namely that individuality itself may be a kind of an illusion, for instance, as expressed by psychologist Charles Fernyhough by the statement that 'we are all fragmented. There is no unitary self. We are all in pieces, struggling to create the illusion of a coherent "me" from moment to moment'.[17] Yet, it is clear that most people exist from day to day with the feeling that they are a unified self and I have highlighted evidence that supports the idea that this a valid feeling. However, for some schizophrenics, a breakdown of conscious awareness may express itself through an inability to recognise that voices that make up the inner speech within our heads are merely expressions of our individuality and not another person somehow existing inside our heads.

How might such an explanation of the schizophrenic state lead to improvements in the diagnosis and treatment of this disorder? In practical terms, brain scans of schizophrenic individuals engaged upon some problem-solving activity that involves conceptualisation might make it possible to see whether coordination of different brain regions during such activity is different compared to that in unaffected individuals. Such studies might identify distinctive dynamic features of the brains of schizophrenic individuals and also highlight differences between such individuals. After all, while a particular range of symptoms define a schizophrenic – symptoms that can include hearing internal voices, experiencing hallucinations, assigning unusual significance or meaning to normal events and experiencing delusions – two people can be diagnosed as suffering from this condition while having a completely different set of symptoms in this range.[18] In other words, what we define as schizophrenia may be a spectrum of loosely related conditions and brain imaging might help define how this relates to different brain dynamics of people in such a spectrum.

Such an approach might also be used to investigate the brain dynamics of depressive individuals. Some types of depression can be associated with delusions, but studies indicate that depression is not associated with the breakdown of conscious awareness found with severe schizophrenia.[19] Therefore, delusions in a depressive individual may have a different root, possibly linked to the feelings of worthlessness characteristic of depression, and such differences might be identified by imaging. Such analyses may also help to define different types of depression, since the condition is increasingly recognised as being as heterogeneous as schizophrenia in its symptoms and perhaps in its biological mechanisms, as well as in the societal pressures that may push an individual into this state.

Ultimately, we need to understand better the mechanisms of the action of drugs that target mental disorders like schizophrenia and depression. This could then provide a platform for developing more specific and effective drugs, with fewer side effects. However, this will require a proper understanding of how such drugs affect the detailed structure and functional responses of the human brain, not vague suggestions based on crude views of the brain as a vat of chemicals. In particular, this understanding will best come from a recognition that human consciousness is mediated by cultural factors, primarily but not only language, and that while the human brain may share similarities with those of other animals, it also has unique differences that affect its function, and how it is influenced by drugs. If such drugs can be developed, they will have much to offer, since even if a mental disorder has a strong environmental component and is rooted in specific social interactions of an individual, drugs can still have an important role to play in treatments. Importantly, drugs should not be seen as an alternative to psychotherapy, but rather as just another tool in a programme in which multiple approaches are tailored to meet an individual's needs.

The more we learn about mental disorders like schizophrenia and depression, the more we are realising that they seem to be linked to a broad spectrum of biological and environmental triggers, but they are also highly varied in their effects on the individual. And while we should be trying to develop a better understanding of the biological basis of different mental conditions, we should also be aware of how dysfunctionality within human society may both be a cause of mental dysfunction and also affect the way we view mental diversity. Ultimately, we should be open to the possibility of changing society for the better. In the next, penultimate, chapter, I will look at how social change can take both positive and negative directions, and how this is linked to changes in human consciousness.

HOW IDEAS CHANGE 19

We live in challenging times. Although civilisation has brought us the marvels of modern technology, music, art and literature, this type of society is also characterised by deep divisions based on class, sexism, racism, war and environmental destruction, all of which can be very detrimental to mental health. Yet it has also been the case that from the birth of civilisation, people have at times organised to resist exploitation and oppression, and fight for a better society. I now want to look at how this might be linked to changes in the minds of those involved, with major changes in the consciousness of humanity as a whole.

A survey of past social movements provides clues as to how human consciousness develops in such circumstances. One is that a sustained protest may begin about a specific matter but can end up addressing much wider issues in society. This is shown by the first recorded example of strike action. This occurred in Egypt in 1155 BC, during the rule of Ramesses III, when skilled artisans who worked on the Egyptian rulers' tombs – the pyramids and other monuments that today draw visitors from across the world – downed tools because of delays in receiving the wheat rations that were the payment for their work.[1] After a prolonged period of action, the artisans were eventually granted all their demands.

The artisans initially stopped work due to concerns about delayed wheat rations, but as the dispute progressed, they began to raise much wider questions about the nature of ancient Egyptian society.[2] A key ideological principle at this time was *ma'at* – a concept of universal, communal and personal balance in society, so that the latter could function in accordance with the will of the gods. While artisans refusing to work might be seen as violating the principle of *ma'at*, the strikers turned this argument on its head by arguing that they had been forced into their actions by a serious breach of this principle, namely a failure of state officials to deliver the workers' rations in a timely manner.

A more recent example of a strike is the year-long one by British miners in 1984–5, precipitated by Prime Minister Margaret Thatcher's decision to close down a substantial section of the UK mining industry.[3] As the strike progressed, the miners began to realise that much of the establishment – from the police to many media outlets – was being marshalled against them. This led to changing perceptions in many miners' minds about their relationship with their local policeman or the newspaper they were used to buying each day.[4] Another change was that, before the strike, the mining community was largely defined by traditional values, with miner's wives expected to have their husband's dinner on the table when he returned from work, despite often having a full-time job themselves. Yet during the strike, through organisations like Women Against Pit Closures, miners' wives and other female supporters played a highly active role and this affected their mindset.[5]

Transformed minds

As one miner's wife, Marie Collins, recalled, 'For the first time the women had some control over their own lives. They weren't just appendages of men – their views mattered. You could see women's confidence growing. They began to challenge the men, to go picketing themselves ... I went on speaking tours around Kent and Scotland and even Germany.'[6] Perhaps the most unexpected pairing during the strike was between the miners and the LGBT+ community.[7] The organisation Lesbians and Gays Support the Miners fundraised in gay pubs and clubs in cities and took the proceeds to mining districts. The experience challenged homophobic prejudice in the coalfields and increased understanding of the strike in the LGBT+ community. Miners returned the solidarity by marching with their union banners on the 1985 Lesbian and Gay Pride demonstration in London.

Instances like these raise questions about what might be happening in the brains of individuals involved in social movements. In this respect it is worth recalling Valentin Voloshinov's view of human consciousness as a boundary phenomenon, with origins in social discourse but situated in the individual mind, and also being a kind of dialogue. As Voloshinov put it, 'Meaning does not reside in the word or the soul of the speaker or the soul of the listener. Meaning is the effect of interaction between speaker and listener produced via the material of a particular sound complex. It is like an electric spark ... Only the current of verbal intercourse endows a word with the light of meaning.'[8] I also mentioned how Voloshinov saw 'speech genres' as imparting a regularity to human communication and also temporarily crystallising relationships between two speakers in terms of their respective power and status; yet speech genres also remain open to the shifting pressures of daily life, as well as being influenced by social and political change.

All of this has relevance for trying to understand what might be changing in the consciousness of some individuals during types of great social upheaval. For there is evidence that social interactions that occur outside normal experience may have transformative effects on individual consciousness. In the 1984–5 strike, miners' wives were initially situated in the home, but through involvement in the support networks that helped to sustain the strike, theirs became a collective experience and brought them into contact with much wider groups of people than would normally be the case. This in turn created a space in which some women began to contest assumptions about their place in society and also raised questions in their minds about issues like gender and sexuality. It would therefore be interesting to study how such changes were accompanied by alterations in the words and the intonation of women who underwent a transformation during the strike.

Successful revolutions – events by which a society's whole structure is transformed – happen rarely, but when they do their consequences are profound, so a common term used to describe the English revolution of 1642–9 was 'the world turned upside down'.[9] A sense of exhilaration is also common in the initial stages of revolutions. The English poet William Wordsworth expressed the hopes of a generation about the French revolution of 1789 when he wrote: 'Bliss was it in that dawn to be alive. But to be young was very heaven!'[10]

Mental progress

Such views reflect the fast-moving pace of revolutions, the way they can challenge long-held views about society and human relationships, but also the unique space they offer to younger people to make their mark on the world. I have already mentioned how the frontal areas of the brain, which are particularly involved in planning, reasoning and judgement, are still maturing in young people, and this may be one reason that individuals at this age can make decisions that seem guided more by emotion than cold rationality – behaviour that is generally considered a sign of immaturity.[11] However, in a revolution, it might mean younger people are far less likely to be influenced by what Karl Marx called the- 'tradition of all dead generations [which] weighs like a nightmare on the brains of the living'.[12]

To a much greater degree than other types of social movement, a central feature of revolutions is their rapid progression from specific and limited demands to ones that can challenge the whole structure of society.[13] So how is such a change, from the specific to the general, reflected in individual human minds? To address this question, we need to return to the notion of thought as a dialogue. Unlike a conversation between two individuals, the inner dialogue that drives an individual consciousness is far less defined in its structure and meaning. We also need to consider the role that the unconscious may play in the human thought process, and how this might be enhanced during periods of upheaval, as well as what we have learned about the often contradictory nature of human consciousness.

One consequence of a contradictory consciousness is that the majority of people in a revolution may initially be reluctant to 'go too far' in their demands, and thereby to look to leaders who espouse a more limited change in society.[154] However, as a revolution progresses and throws up deeply rooted dilemmas and contradictions, this can lead to a more questioning attitude among some people and an increased willingness to follow leaders of a more radical persuasion. This can explain how individuals who are initially minor players in society can come to play a much more prominent role as a revolution unfolds.

For instance, in the 1642–9 revolution that ended feudalism in England, Oliver Cromwell was initially a marginal figure. However, his determination to defeat the Royalist forces, as opposed to the half-hearted efforts of many other Parliamentary leaders, brought him to prominence. This determination led Cromwell to select officers based on ability and willingness to wage the war as ruthlessly as required, rather than on their perceived social status. As Cromwell himself expressed it: 'I had rather have a plain, russet-coated Captain, that knows what he fights for, and loves what he knows, than that which you call a Gentleman and is nothing else.'[14] Such a new approach radicalised the constitution of the army, but itself reflected a radicalisation taking place in English society in general. For a short while, this led to a very radical group indeed, the Levellers, coming to prominence.[15]

Two common viewpoints exist regarding revolutions. One is that they are purely spontaneous affairs; another is that they are the product of a few conspirators with the masses as mere spectators. Yet neither viewpoint adequately reflects the reality of how genuine revolutions progress. Regarding spontaneity, it is true that revolutionary situations can progress very rapidly, reflecting a radicalisation among the mass of people. This is one reason why revolutions often seem to come out of the blue, surprising even many would-be revolutionaries. The Russian revolution of February 1917 began with a demonstration led by women campaigning for more bread for their families.[16] Initially, the women were encouraged to return home by local Bolshevik male leaders, who feared the demonstration might be used as an excuse by the authorities for a crackdown. However, the women ignored this advice and kick-started a movement that ended the centuries-old Tsarist monarchy.

Important as spontaneity is, successful revolutions have also always involved strong political leadership. A key aspect in the radicalisation of the English revolution of 1642–9 was the agitation carried out by the Levellers. Through the use of petitions, pamphlets and public demonstrations, this group built a movement that influenced both members of the army and the public as a whole.[17] The Bolsheviks only gained a majority in the 'soviets' – the elected bodies that became an organising forum for revolutionary workers and peasants – after months of agitating in the factories and streets.[18] Having achieved such a majority, the Bolsheviks used the soviets to mount a challenge for control of the state.

If mass social movements involve major changes in the ways people perceive what is acceptable, and possible, in society, how might this relate to brain function? Two parts of the brain that I have already mentioned – the amygdala and the prefrontal cortex – are important in this context. As befits our status as rational beings, the prefrontal cortex plays a particularly important role in humans in terms of our interactions with others. Yet there is increasing recognition that a lot of what we take for granted about the state of the world – and this can include prejudices about others based on their skin colour, nationality, sexual persuasion or gender – also involves unconscious feelings, and the amygdala plays an important role in such feelings.[19] In fact, the interaction between the prefrontal cortex and amygdala is dynamic and in a situation of rapid social change, as in a mass movement, I believe there is the potential for the prefrontal cortex to play a much more dominant role.

The reason for this is that mass movements force individuals into unfamiliar situations and lead to new types of social interactions with other people in which language plays a key role. This restructuring of neural impulses in the prefrontal cortex could then start to challenge the type of 'unconscious biases' that are rooted in the amygdala. In fact, recent studies have shown that the prefrontal cortex itself is far from homogenous in this respect. Instead, such studies have shown that while the ventromedial prefrontal cortex is involved in guiding our future actions based on things that have happened to us in the past, in contrast, the dorsomedial part of this brain region allows us to adjust our behaviour to a rapidly changing situation.[20] This makes evolutionary sense for while past experiences may be a good guide to action in a stable environment, they may not remain so in times of uncertainty. Therefore, the dorsomedial prefrontal cortex is likely to play a more dominant role during the rapidly unfolding events characteristic of major social upheavals.

The examples that I have given of revolutions may all appear to be from the distant past, so it is worth mentioning a more recent example – the Egyptian revolution of 2011. A retired businessman who travelled from Alexandria to join the protesters in Cairo gathered to demand the fall of dictator Hosni Mubarak, noted: 'There were all kinds of people. From universities, secondary schools, preparatory schools. Homeless people. People from every religion. All divisions disappeared. Everyone had one purpose. I was really crying, for this was the first time I saw the Egyptian people unafraid of anything.'[21] Sadly, this lack of fear was very brief, for although Mubarak was indeed deposed by the mass movement, and a democratically elected leader – Mohamed Morsi – was installed, a new military dictatorship soon took power. Subsequently, there has been a major crackdown by the new regime on human rights, particularly against women, LGBT+ individuals and religious minorities.[22]

Resisting reaction

A similar reaction took place in the Soviet Union in the 1930s under Josef Stalin, with women's rights replaced by a celebration of motherhood and the family. And despite Jews having played important roles in the 1917 revolution and gained civil rights in that revolution, now once more they became a scapegoat for society's troubles. Similarly, having championed the right of nations to self-determination, the Soviet Union under Stalin reverted to the same 'prison house of nations' that it had been under the Russian Empire of the Tsar.

Ironically, despite their claims to represent opposite poles of the political spectrum, both the Soviet Union under Stalin and the German Nazi regime of Adolf Hitler were remarkably similar in their reactionary attitudes to women, LGBT+ individuals and ethnic and religious minorities. Perhaps this is not so surprising, given that both regimes were based on a totalitarian system that moved quickly and ruthlessly to quash any dissent. Obviously one factor to consider here is the potential risk in voicing opposition to a dictatorship such as the ones led by Hitler or Stalin, but it is also worth asking what psychological explanations there might be for such apparent mass acquiescence to evil.

One study that has strongly influenced our thinking in this area was carried out by the psychologist Stanley Milgram in the 1960s.[23] Milgram wanted to test the claim by the German Nazi Adolf Eichmann, during his war crimes trial, that he and his accomplices in the Holocaust were 'just following orders'. In the study, volunteers were asked to 'electrocute' other human 'learners', having been told that this was a necessary part of the study. Although the learners appeared to scream in pain, in reality, they were only acting and there was no electric shock. However, the volunteers did not know this and, shockingly, the majority of them showed a willingness to do as they had been instructed and electrocute the learners.

On the face of it, this study seems to confirm the idea that most people will acquiesce with authoritarianism and evil. Yet although troubling, there is surely a danger in extrapolating from a highly artificial laboratory study, without also surveying the actual ways that individuals in history have reacted to real-life authoritarian regimes like that of Hitler or Stalin. What such a survey shows is acquiescence by many, but also resistance and the greatest bravery by some individuals, even in the most difficult of circumstances.

For instance, on a recent visit to Munich, I was deeply moved by the monument to Sophie and Hans Scholl, siblings who were part of the White Rose student movement that secretly distributed anti-war leaflets in cities across Germany from 1942 onwards. Captured by the Nazis in February 1943, the Scholls were tried and swiftly executed. Just before her death, Sophie said: 'Such a fine, sunny day, and I have to go, but what does my death matter, if through us thousands of people are awakened and stirred to action?' I could also cite other examples, such as Oskar Schindler, the industrialist and Nazi party member who nevertheless saved more than 1,000 Jewish individuals from the death camps.

Such examples confirm that during progressive social movements, but also at the heights of deepest reaction, individual actions can make a tremendous difference to events. We therefore need to explain in terms of brain function how it is that some individuals react to such social events in a progressive fashion, yet others acquiesce with, or even support, reaction. Given what I said earlier about the likely changing interaction between the amygdala and the prefrontal cortex in times of social upheaval, there may be some fruitful future research to be done in terms of studying how this interaction differs between individuals who acquiesce and those who rebel, during specific social upheavals.

FUTURE OF MIND $\quad$ 20

The material basis of human consciousness is one of the biggest unsolved issues in science. It would therefore hardly be surprising for any book about this subject to do anything more than scratch the surface of this vast and complex topic. Nevertheless, I hope that this book has been revealing in showing how modern neuroscience and psychology are expanding our understanding of the biological and social mechanisms that underlie human consciousness, and also pointing to ways that we can increase such understanding through future studies.

As I mentioned at the start of this book, the philosopher René Descartes revolutionised our understanding of how the human body works by proposing that it can be understood as a machine, yet he failed to extend his mechanistic view to the mind. Descartes argued instead that 'higher' mental functions can only be explained by reference to an immortal soul, opaque to scientific investigation. This introduced a dualism that persists to this day in discussions about human consciousness. We therefore need to consider how much the view of the human mind developed here both recognises and overcomes this dualism.

The difficulties in developing a view of human consciousness that does not end up as Cartesian dualism arise from two main directions. The first comes from the difficulty in accepting that something as personal as the individual 'I' is merely based on the unconscious actions of cells in the brain, particularly given the apparent complexity of human thought that at times seems to verge on the transcendental. The second difficulty arises from the problem of how to construct a view of the mind that can explain our sense of being an individual consciousness, without lapsing into a scenario in which that consciousness appears to be due to a tiny self-aware homunculus sitting inside our heads.

In fact, I believe that the view of human consciousness developed in this book can explain the uniqueness of our species' conscious awareness, but in an entirely materialistic fashion. This approach views language – the system of abstract symbols linked in a grammatical structure but also one that connects the individual to the world outside via word meaning – as a material force that has reshaped human brain functions at every level. This has led to a qualitative shift in such functions, compared to that of every other species.

Mental evolution

Notably, this view of the mind sees the different brain regions, and their interconnections, as substantially altered in humans. Throughout this book, I have identified examples of such changes. For instance, we have evidence that the cerebellum – formerly thought to be only involved in repetitive motor skills – has expanded in size in humans and increased its connections with the cerebral cortex, allowing it to play unique roles in human imagination and creativity. Other changes include the expression of certain genes in different brain regions becoming altered in humans. For example, the gene TH, involved in secretion of the neurotransmitter dopamine, is expressed in the human brain not only in the striatum – a region involved in movement – as in other species, but also in the neocortex, pointing to distinctive roles for dopamine in 'higher' human brain functions.

Although such changes point to an altered structural architecture of the human brain, another important feature of our species is that our brains are far more 'plastic' than those of other species, including our closest biological cousins, the great apes. This plasticity is particularly important for the development of each individual human consciousness, because although human beings are biologically primed to develop our unique form of self-conscious awareness, this only happens if we are exposed to the richness of human society. Exposure to such a society, particularly through language, but also through other 'cultural tools' and technologies, is a key aspect of how an individual human consciousness is formed.

Importantly, this means that human consciousness is as much a social as an individual entity, with Valentin Voloshinov seeing it as a 'boundary phenomenon'. Somehow we must explain how its social aspects can nevertheless exist in a biological organ that sits within the bony confines of an individual's skull. We also need to explain how that biological brain is connected to the sensory organs and other parts of a person's body.

Because language has transformed human consciousness, we can gain valuable insights into the dynamics of such consciousness by studying, and speculating about, the ways in which 'inner speech' and other symbolic forms influence brain processes. Ultimately, however, we need to explain human consciousness in terms of molecular and cellular mechanisms in the brain. That means engaging with the latest insights from neuroscience. In particular, although we can make some general points about the material basis of human consciousness from what neuroscience is revealing about the architecture of the brain, and the biological differences identified between human brains and those of other species, what we really want is insights into human brain dynamics. I believe that the most exciting recent development is what we are learning about brain waves.

I have highlighted how brain waves play an important role in coordinating different brain regions in particular activities that include working memory. In fact, it seems very likely that brain waves are key to understanding consciousness in various species besides humans. It is likely that such waves play a coordinating role when your pet dog or cat suddenly notices something novel in its environment, whether a squirrel, a tasty snack or just the sight of you. However, in humans, superimposed on this mechanism for switching attention from one thing to another, language also must play a unique and higher guiding role – acting through brain waves, but doing so in a structured way that can only occur in our species – because of the word meaning and grammatical structure that language alone provides. Investigating how language affects the coordinating role of different brain waves associated with various aspects of human consciousness is likely to be a fertile area for future studies.

Complex consciousness

Although studies highlighting the important role of brain waves help us to understand how incoming information is processed within the brain and how different brain regions interact on a moment-to-moment level, we should not forget that there is a lot more to brain function than electrical activity. It is increasingly recognised that not only electrical communication but also that mediated by chemicals, such as neurotransmitters and by newly discovered 'signalling' molecules like regulatory RNAs, play important roles in the brain. It will therefore be important in future to study how these other forms of communication might be involved in reconfiguring the human brain in response to language and other cultural mediators, particularly in terms of long-term changes in the brain.

One question to ask is whether the view of consciousness presented in this book goes any way to address the 'hard' problem of consciousness first outlined by the philosopher David Chalmers. Recall that Chalmers argued that although modern neuroscience might soon provide many insights into the nature of a variety of thought processes, he was more sceptical about the idea that we would ever be able to explain, in material scientific terms, that subjective sense we have as individuals of being us, with all the richness of thought and experience that is associated with such subjectivity.

In some ways I believe I have addressed the 'hard' problem in this book and hopefully demonstrated that there is nothing magical about human consciousness. Instead, it is becoming clear that not only are we gaining a better understanding of the molecular and cellular basis of brain function, but the new insights about the role of different frequency brain waves provides a material basis for the dynamics of thought. Of course, I have also said that language has transformed human consciousness in ways that make it unique compared to that of other species. However, since words can also be considered as material forces, at least in the context of brain function, this added aspect of human consciousness can also be explained in a materialist way, with no requirement for a 'soul' or explanations that go beyond the known laws of science, as is the case for panpsychist explanations for the mind.

What about the difficulty in ever knowing what it would feel like to experience things through the eyes of another human being? I think one important thing to say here is that increasing insights about the genes involved in perception and subtle differences between people in such genes, combined with the very different social experiences that make up a person's lifetime, mean that different individuals may experience the same events in very different ways. However, that does not mean that explaining such subjectivity in material terms will be impossible in the future, but it would be very complicated. Another important point to make is that we should not be relying on science alone to explore such subjectivity.

Despite the obvious difficulties in trying to 'get inside' another person's head, one important consequence of the fact that – uniquely among the species on planet Earth – human beings can communicate with other humans through speech, and also literature, art, music and film, means that we can convey some of what it feels like to be 'us' in ways that involve both conscious and unconscious thoughts. Because of this, I would argue that it is indeed possible to access another person's mind, through reading a novel, seeing a work of art, hearing a piece of music or watching a film. Depending on the quality of the artform, we may not only gain insights into the author's mind, but ones that touch on the deepest contradictions of human society, and perhaps even what it means to be human.

If gaining insights into the meaning of life provides one reason for studying human consciousness and its connection to brain function, another reason is to better understand psychiatric disorders like schizophrenia and clinical depression. So what impact, if any, might the approach to understanding human consciousness that I have been developing in this book have upon diagnosis and treatment of such disorders? We should first acknowledge how complex a phenomenon mental disorder is, both in biological and social terms. There are unlikely to be easy answers for those seeking to reduce human mental disorders to either genetics or societal pressures. However, this does not mean that the view of human consciousness advanced here has nothing to offer in terms of better diagnosis and treatment, and indeed, I have highlighted some of the ways an increased understanding may lead to new forms of diagnosis and treatment for various mental disorders and conditions. Psychotherapy has an important role to play as well as the development of new drugs, but any treatment needs to consider the patient as a key player in such treatment.

Unstable species

Something we should consider is whether the human mind is particularly vulnerable to mental disorder as a by-product of our evolution. For instance, the psychiatrist Ole Andreassen has recently found evidence that genetic differences that distinguish humans from Neanderthals may have enhanced our powers of creativity and imagination, but also made us more vulnerable to schizophrenia. As well as showing that genetic differences linked to this disorder were only found in the genomes of modern humans, and not in Neanderthals, findings by Andreassen indicate that the brain regions in which the genes are expressed are linked to our human ability to think and understand in complex ways.

However, while vulnerability to mental disorder may be the unfortunate price we pay for our conscious awareness, it would be a major oversight not to also seek explanations for the current epidemic of mental disorder in existing society. This society has given us immense power, as our cities and technologies increasingly reach into every corner of the globe, and even across the solar system. Proof of our species' ability to sustain itself is demonstrated by the fact that the human population now numbers 8 billion and has grown more since the 18th century than in the entire history of humanity prior to that moment. Yet while current society has brought us civilisation, and provided great material prosperity and wealth for humanity as a whole, it is also characterised by huge levels of inequality, both globally and within nations, and increasing economic and environmental crises that can be destabilising to individual human minds and may ultimately threaten the future of human civilisation itself, with all the implications that may have for mental health in the future.

Given such major anxieties and societal pressures, it is not surprising that many individuals today feel a sense of despair for the future – not just in terms of their own immediate futures and those of their families and friends – but about the long-term prospects for humanity and more generally life on our planet. This feeling was only exacerbated by the global Covid-19 pandemic that swept the world in 2020. This led not only to many deaths but also a lockdown that brought the global economy to a juddering halt and confined individuals to their homes, cut off from physical contact with many family members and friends. More recently, the war in Ukraine has led to new fears about the potential for this turning into a nuclear war. Human activities are also leading to the mass extinction of species, with one in four mammals currently at risk of going extinct and 41 per cent of amphibians. However, undoubtedly the biggest challenge of all for humanity is global warming, driven by human-generated carbon emissions, which threatens not only human civilisation but possibly the future of life on Earth. Despite regular international political summits that end with promises to tackle the problem of global warming, in practice little is being done, with carbon emissions actually increasing over recent years.

All of this sounds like grounds for pessimism. However, I believe we can also find good reasons to be positive about our current situation. One positive development is the innovative approaches and technologies that young people in particular are bringing to social movements. Not that this should be surprising. After all, a key feature of our species is that the brain is especially plastic in human beings, even into their late twenties, and as a consequence younger people are always more open to new ways of doing things. Yet even after our formative years, human brains remain far more plastic than had been realised, meaning that each one of us has the potential for our thoughts and assumptions to change. At the same time, it is increasingly recognised that certain individuals whose brains are 'atypical' have much to offer to society in being able to see beyond what is accepted as normal and inevitable, yet which may be detrimental to humanity's long-term well-being and the future of the planet. Greta Thunberg, who has become a symbol of the global movement to 'save the planet', has explicitly linked her stance on this question to her autism.[1]

A central theme of this book is that human consciousness differs from that of other species in that only humans can conceptualise the world around us through language. This allows us to learn lessons from the past and imagine different types of our world in the future. Coupled with our other unique gift, the ability to design and use novel tools and technologies in each new human generation, this gives our species the power to shape the world around us in a controlled, planned way. This is our potential, yet currently we seem to have relinquished that control as our activities threaten to destroy the environment and our civilisation with it. Yet the potential remains and, despite daunting obstacles ahead, collectively we have the powers of creativity and imagination that could allow us to build a very different sort of society, one that is sustainable and works not against nature but in harmony with it, and in which technology enriches our lives, rather than oppressing us. If we can develop such a civilisation, who knows what marvels we might achieve in the future, not only on Earth, but also on other planets. If so, future generations may marvel that such developments have emerged from the human brain, an unprepossessing object weighing only 1.5 kilograms, and with the appearance and consistency of cold porridge.

REFERENCES

1 What is Consciousness?

1. Mesaroş, C. Aristotle and animal mind. *Procedia – Social and Behavioral Sciences*, 163, 185–192 (2014).
2. Cunning, D. Descartes on the Dubitability of the Existence of Self. *Philosophy and Phenomenological Research*, 74, 111–131 (2007).
3. Barnes, J. How the dualism of Descartes ruined our mental health. *Aeon* (10 May 2019).
4. Inukai, Y. Hume's Labyrinth: The Bundling Problem. *History of Philosophy Quarterly*, 24, 255–274 (2007).
5. Lodge, P.B., Bobro, M. Stepping Back Inside Leibniz's Mill. *The Monist*, 81, 553–572 (1998).
6. Chalmers, D.J. Facing Up to the Problem of Consciousness. *Journal of Consciousness Studies*, 2, 200–219 (1995).
7. Papineau, D. Materialism must be defended. *Institute of Art and Ideas* (12 March 2020).
8. Johnson, G. What Really Goes On in There. *The New York Times* (10 November 1991).
9. Ritchie, S. Do Daniel C. Dennett's memes deserve to survive? *The Spectator* (4 March 2017).
10. Dawkins, R. *The Selfish Gene: 30th Anniversary edition*, p. 192 (Oxford University Press, 2006).
11. Kuhn, A. He Helped Discover Evolution, And Then Became Extinct. *NPR* (30 April 2013).

12. Reese, B. Interview with Christof Koch. *GigaOm* (25 May 2018).
13. Nagel, T. What Is It Like to Be a Bat? *The Philosophical Review*, 83, 435–450 (1974).

2 Tools and Symbols
1. Engels, F. The Part played by Labour in the Transition from Ape to Man. https://www.marxists.org/archive/marx/works/1876/part-played-labour/ (1876).
2. Trigger, B. Comment on Tobias, Piltdown, the case against Keith. *Current Anthropology*, 33, 274–275 (1992).
3. Hawks, J. Human evolution is more a muddy delta than a branching tree. *Aeon* (8 February 2016).
4. Hu, J.C. What Do Talking Apes Really Tell Us? *Slate* (21 August 2014).
5. Jabr, F. Catching Ourselves in the Act of Thinking. *Scientific American* (15 October 2013).
6. Lourenço, O.M. Piaget and Vygotsky: Many resemblances, and a crucial difference. *New Ideas in Psychology*, 30, 281–295 (2012).
7. Beck, J. The Running Conversation in Your Head. *The Atlantic* (23 November 2016).
8. Oakes, K. What the voice inside your head says about you. *BBC Future* (20 August 2019).
9. Fernyhough, C. Getting Vygotskian about theory of mind: Mediation, dialogue, and the development of social understanding. *Developmental Review*, 28, 225–262 (2008).

3 Nerves and Brains
1. Ribatti, D. William Harvey and the discovery of the circulation of the blood. *Journal of Angiogenesis Research*, 1, 3 (2009).
2. Kwon, D. New Discoveries in Human Anatomy. *The Scientist* (18 February 2020).
3. Newman, T. All you need to know about neurons. *Medical News Today* (7 December 2017).
4. Elam, J.S. et al. The Human Connectome Project: A retrospective. *Neuroimage*, 244, 118543 (2021).
5. MICrONS Consortium et al. Functional connectomics spanning multiple areas of mouse visual cortex. *bioRxiv* (29 July 2021).
6. Jäkell, S., Dimou, L. Glial Cells and Their Function in the Adult Brain: A Journey through the History of Their Ablation. *Frontiers in Cellular Neuroscience*, 11, 24 (2017).

7. Hunter, P. The paradox of model organisms. The use of model organisms in research will continue despite their shortcomings. *EMBO Reports*, 9, 717–720 (2008).

8. Azkona, G., Sanchez-Pernaute, R. Mice in translational neuroscience: What R we doing? *Progress in Neurobiology*, 217, 102330 (2022).

9. Fernandez, E. Human Participants In Experimentation On The Brain? They Better Be Treated Well. *Forbes* (21 November 2019).

10. What they said: Genome in quotes. BBC News (26 June 2000).

11. Donahue, M.Z. New Clues to How Neanderthal Genes Affect Your Health. *National Geographic* (5 October 2017).

12. Fiddes, I.T. et al. Human-Specific NOTCH2NL Genes Affect Notch Signaling and Cortical Neurogenesis. *Cell*, 173, 1356–1369 (2018).

4 Evolving Minds

1. Gleiser, M. Does life on Earth have a purpose? *Big Think* (16 November 2022).

2. Bi, S. & Sourjik, V. Stimulus sensing and signal processing in bacterial chemotaxis. *Current Opinion in Microbiology*, 45, 22–29 (2018).

3. Staughton, J. How Long Did It Take For Multicellular Life To Evolve From Unicellular Life? *Science ABC* (6 January 2022).

4. Satterlie, R.A. Do jellyfish have central nervous systems? *Journal of Experimental Biology*, 214, 1215–1223 (2011).

5. Robson, D. A brief history of the brain. *New Scientist* (21 September 2011).

6. Bailey, R. Divisions of the Brain. *ThoughtCo* (15 November 2019).

7. Defelipe, J. The evolution of the brain, the human nature of cortical circuits, and intellectual creativity. *Frontiers in Neuroanatomy*, 5, 29 (2011).

8. Bailey, R. The Limbic System of the Brain, *ThoughtCo* (28 March 2018).

9. Konnikova, M. The man who couldn't speak and how he revolutionized psychology. *Scientific American* (8 February 2013).

10. Cherry, K. What is Wernicke's Area? *Very Well Mind* (26 November 2022).

11. Gold, J. A WHAT Lives Inside My Brain? *NPR* (18 March 2009).

12. Sanders, L. Brain waves may focus attention and keep information flowing. *Science News* (13 March 2018).

13. Hrvoj-Mihic, B., Bienvenu, T., Stefanacci, L., Muotri, A.R., Semendeferi, K. Evolution, development, and plasticity of the human brain: from molecules to bones. *Frontiers in Human Neuroscience* 7, 707 (2013).

14. Eckart-Washington, K. Brain differences in blind people may sharpen hearing. *Futurity* (6 May 2019).
15. Gregoire, C. Research Uncovers How And Where Imagination Occurs In The Brain. *Huffington Post* (18 September 2013).
16. Costandi, M. Human Brain Cells Boost Mouse Memory. *Science* (7 March 2013).
17. Bergland, C. What Makes Us Human? Dopamine and the Cerebellum Hold Clues. *Psychology Today* (24 November 2017).

5 Thought and Reason

1. Nunn, G. Why do pigs oink in English, boo boo in Japanese, and nöff-nöff in Swedish? *The Guardian* (18 November 2014).
2. Monaghan, P., Shillcock, R.C., Christiansen, M.H., Kirby, S. How arbitrary is language? *Philosophical Transactions Royal Society London B Biological Sciences*, 369, 20130299 (2014).
3. Nowak, M.A., Krakauer, D.C. The evolution of language. *Proceedings of National Academy of Science USA*, 96, 8028–8033 (1999).
4. Terrace, H.S. Why Chimpanzees Can't Learn Language 1. *Psychology Today* (2 October 2019).
5. Bickerton, D. Pidgin and Creole Studies. *Annual Review of Anthropology*, 5, 169–193 (1976).
6. Sessarego, S. Not all grammatical features are robustly transmitted during the emergence of creoles. *Humanities and Social Sciences Communications*, 7, 130 (2020).
7. Popovich, N. How the Deaf Brain Rewires Itself to 'Hear' Touch and Sight. *The Atlantic* (11 July 2012).
8. Timmerman N., Ostertag, J. Too Many Monkeys Jumping in Their Heads: Animal Lessons within Young Children's Media. *Canadian Journal of Environmental Education*, 16, 59–75 (2011).
9. Choi, C.Q. Human Evolution: The Origin of Tool Use. *LiveScience* (11 November 2009).
10. Munkittrick, K. Did Humans Make Tools, or Did Tools Make Humans? *Discover Magazine* (26 August 2010).

6 The Sensual World

1. Williams, D.L. Light and the evolution of vision. *Eye*, 30, 173–178 (2016).
2. Shalaeva, D.N., Galperin, M. Y., Mulkidjanian, A.Y. Eukaryotic G protein-coupled receptors as descendants of prokaryotic sodium-translocating rhodopsins, *Biology Direct*, 10, 63 (2015).

3. Grady, D. The Vision Thing: Mainly in the Brain. *Discover Magazine* (31 May 1993).

4. Kean, S. The Cat Nobel Prize Part I. *Psychology Today* (14 April 2014).

5. Ku, L.Y. Cats and Vision: is vision acquired or innate? *The Serious Computer Vision Blog* (1 June 2013).

6. Beyeler, M., Rokem, A., Boynton, G.M., Fine, I. Learning to see again: biological constraints on cortical plasticity and the implications for sight restoration technologies. *Journal of Neural Engineering*, 14, 051003 (2017).

7. Sherman, S.M., Guillery, R.W. The role of the thalamus in the flow of information to the cortex. *Philosophical Transactions Royal Society London B Biological Sciences*, 357, 1695–1708 (2002).

8. LoBue, V., Adolph, K.E. Fear in infancy: Lessons from snakes, spiders, heights, and strangers. *Developmental Psychology*, 55, 1889–1907 (2019).

9. Sliney, D.H. What is light? The visible spectrum and beyond. *Eye*, 30, 222–229 (2016).

10. Spector, D., Snodgrass, E. Photos show how cats see the world compared to humans. *Business Insider* (16 October 2013).

11. Brooks, M. Drawing, Visualisation and Young Children's Exploration of 'Big Ideas'. *International Journal of Science Education*, 31, 319–341 (2009).

12. Saito, A., Hayashi, M., Takeshita, H., Matsuzawa, T. The origin of representational drawing: a comparison of human children and chimpanzees. *Child Development*, 85, 2232–2246 (2014).

7 Learning and Memory

1. Bailey, C.H., Kandel, E. R. Synaptic remodeling, synaptic growth and the storage of long-term memory in Aplysia. *Progress in Brain Research*, 169, 179–198 (2008).

2. Marx, G., Gilon, C. The Molecular Basis of Memory. *ACS Chemical Neuroscience*, 3, 633–642 (2012).

3. Akst, J. Glial cells aid memory formation. *The Scientist* (12 January 2010).

4. Shapin, S. The Man Who Forgot Everything. *The New Yorker* (14 October 2013).

5. Deisseroth, K. Optogenetics: Controlling the Brain with Light. *Scientific American* (20 October 2010).

6. Trafton, A. Neuroscientists identify brain circuit necessary for memory formation. *MIT News* (6 April 2017).

7. Ng, N. 'Heart of brain' breakthrough may aid treatment of disorders, Hong Kong scientists say. *South China Morning Post* (18 September 2017).
8. Goldhill, J. The 'Jennifer Aniston neuron' is the foundation of compelling new memory research. *Quartz Magazine* (23 July 2016).
9. Abbott, A., Callaway, E. Nobel prize for decoding brain's sense of place. *Nature*, 514, 153 (2014).
10. Gosline, A. Why your brain has a Jennifer Aniston cell. *New Scientist* (22 June 2005).

8 Mind Chemistry

1. Dixon, T. 'Emotion': The History of a Keyword in Crisis. *Emotion Review*, 4, 338–344 (2012).
2. Guy-Evans, O. Neurotransmitters: Types, Function and Examples. *Simply Pyschology* (9 February 2023).
3. Pallardy, R. Dogs Have Co-Evolved With Humans Like No Other Species. *Discover Magazine* (3 November 2021).
4. Weymar, M., Schwabe, L. Amygdala and Emotion: The Bright Side of It. *Frontiers in Neuroscience*, 10, 224 (2016).
5. Fineberg, S.K., Ross, D.A. Oxytocin and the Social Brain. *Biological Psychiatry*, 81, e19–e21 (2017).

9 Philosophy of Mind

1. Maden, J. George Berkeley's Subjective Idealism: The World Is In Our Minds. *Philosophy Break* (April 2021).
2. Frankish, K. in *Philosophers on Consciousness: Talking about the Mind* (ed. J. Symes), pp. 89–99 (Bloomsbury Academic, 2022).
3. Strawson, G. The Consciousness Deniers. *The New York Review of Books* (13 March 2018).
4. Koch, C. Is Consciousness Universal? *Scientific American* (1 January 2014).
5. Paulson, S. Roger Penrose On Why Consciousness Does Not Compute. *Nautilus* (27 April 2017).
6. Martin, A.H., K. Emergence: the remarkable simplicity of complexity. *The Conversation* (30 September 2014).
7. Powell, C.S. Is the Universe Conscious? *NBC News* (16 June 2017).

10 Individual and Society

1. Voloshinov, V.N. *Marxism and the Philosophy of Language*, p. 13 (Harvard University Press, 1973.).
2. Orazbekova, Z.S., N., Mamyrova, K., Zhumabayeva, A. The Intonation in Gender Analysis of Linguistics. *European Proceedings of Social and Behavioural Sciences*, 6, 11–20 (2015).
3. Voloshinov, V.N. *Marxism and the Philosophy of Language*, p. 41 (Harvard University Press, 1973).
4. McLeod, S. Pavlov's Dogs. *Simply Psychology* (9 February 2013).
5. McLeod, S. Classical Conditioning. *Simply Psychology* (8 March 2014).
6. Virues-Ortega, J. The Case Against B.F. Skinner 45 years Later: An Encounter with N. Chomsky. *Behavior Analyst*, 29, 243–251 (2006).
7. Guess, D., Sailor, W., Rutherford, G., Baer, D.M. An experimental analysis of linguistic development: the productive use of the plural morpheme. *Journal of Applied Behavioral Analysis*, 1, 297–306 (1968).
8. Palmer, D.C. On Chomsky's Appraisal of Skinner's Verbal Behavior: A Half Century of Misunderstanding. *Behavior Analyst*, 29, 253–267 (2006).
9. Cepelewicz, J. In Birds' Songs, Brains and Genes, He Finds Clues to Speech. *Quanta Magazine* (30 January 2018).

11 Information and Meaning

1. Atkisson, M. Behaviorism vs. Cognitivism. *Ways of Knowing* (12 October 2010).
2. McLeod, S. Information Processing Theory. *Simply Psychology* (26 February 2023).
3. Cobb, M. Why your brain is not a computer. *The Guardian* (27 February 2020).
4. Todesco, M. The secret ultraviolet colours of sunflowers attract pollinators and preserve water. *The Conversation* (21 February 2022).
5. Neitz, M., Neitz, J. Molecular Genetics of Color Vision and Color Vision Defects. *Archives of Ophthalmology*, 118, 691–700 (2000).
6. Breeden, A. When Covid-19 Stole Their Smell, These Experts Lost Much More. *The New York Times* (19 September 2021).

12 Chance and Design

1. Browne, J. Wallace and Darwin. *Current Biology*, 23, R1071–1072 (2013).
2. Koseska, A., Bastiaens, P.I. Cell signaling as a cognitive process. *EMBO Journal*, 36, 568–582 (2017).
3. Gallo, V.P., Accordi, F., Chimenti, C., Civinini, A., Crivellato, E. Catecholaminergic System of Invertebrates: Comparative and Evolutionary Aspects in Comparison With the Octopaminergic System. *International Review of Cell and Molecular Biology*, 322, 363–394 (2016).
4. Ranjbar-Slamloo, Y. Fazlali, Z. Dopamine and Noradrenaline in the Brain; Overlapping or Dissociate Functions? *Frontiers in Molecular Neuroscience*, 12, 334 (2019).
5. Gregory, T.R. The Argument from Design: A Guided Tour of William Paley's Natural Theology (1802) *Evolution: Education and Outreach* 2, 602–611 (2009).
6. Stephenson, A., Adams, J.W. Vaccarezza, M. The vertebrate heart: an evolutionary perspective. *Journal of Anatomy*, 231, 787–797 (2017).

13 Structure and Function

1. Banks, T. Bauhaus: What was it and why is it important today? *Design Week* (18 February 2019).
2. Ogawa, T., de Bold, A.J. The heart as an endocrine organ. *Endocrine Connections*, 3, R31–44 (2014).
3. Simion, F., Giorgio, E.D. Face perception and processing in early infancy: inborn predispositions and developmental changes. *Frontiers in Psychology*, 6, 969 (2015).
4. Naughton, J. How the Madeleine will help with remembrance of smells past. *The Guardian* (26 October 2013).
5. Schacter, D.L. et al. The future of memory: remembering, imagining, and the brain. *Neuron*, 76, 677–694 (2012).
6. Smith, M. Memory: A curious journey. *The Emotional Learner* (23 February 2020).
7. Carlen, M. What constitutes the prefrontal cortex? *Science*, 358, 478–482 (2017).
8. García-Molina, A. Phineas Gage and the enigma of the prefrontal cortex. *Neurología*, 27, 370–375 (2012).
9. Woodward, A. With a Little Help from My Friends. *Scientific American* (1 May 2017).

14 Circuits and Waves

1. Blakemore, E. How Twitching Frog Legs Helped Inspire 'Frankenstein'. *Smithsonian Magazine* (4 December 2015).
2. Jabr, F. Know Your Neurons: The Discovery and Naming of the Neuron. *Scientific American* (14 May 2012).
3. Ormerod, W. Richard Caton (1842–1926): pioneer electrophysiologist and cardiologist. *Journal of Medical Biography*, 14, 30–35 (2006).
4. Zhang, Z., Sun, Q.Q. The balance between excitation and inhibition and functional sensory processing in the somatosensory cortex. *International Review of Neurobiology*, 97, 305–333 (2011).
5. Fields, R.D. Do Brain Waves Conduct Neural Activity Like a Symphony? *Scientific American* (29 November 2018).
6. Miller, E.K. The 'working' of working memory. *Dialogues in Clinical Neuroscience*, 15, 411–418 (2013).
7. Orenstein, D. How working memory gets you through the day. *MIT News* (24 October 2018).
8. Cromer, J.A., Roy, J.E., Miller, E.K. Representation of multiple, independent categories in the primate prefrontal cortex. *Neuron*, 66, 796–807 (2010).
9. Miller, E.K. Comment on Twitter. https://twitter.com/millerlabmit/status/1108082822026190848?lang=ar (19 March 2019).
10. Fries, P.A mechanism for cognitive dynamics: neuronal communication through neuronal coherence. *Trends in Cognitive Science*, 9, 474–480 (2005).
11. Miller, E.K., Lundqvist, M., Bastos, A.M. Working Memory 2.0. *Neuron*, 100, 463–475 (2018).
12. Fries, P. Rhythms for Cognition: Communication through Coherence. *Neuron*, 88, 220–235 (2015).

15 Free Will and Selfhood

1. Hayry, M. The tension between self governance and absolute inner worth in Kant's moral philosophy. *Journal of Medical Ethics*, 31, 645–647 (2005).
2. Frankish, K. The demystification of consciousness. *Institute of Art and Ideas* (20 March 2020).
3. Buckley, A. Is consciousness just an illusion? *BBC News* (4 April 2017).
4. Wegner, D.M., Wheatley, T. Apparent mental causation. Sources of the experience of will. *American Psychologist*, 54, 480–492 (1999).

5. Klemm, W.R. Free will debates: Simple experiments are not so simple. *Advances in Cognitive Psychology*, 6, 47–65 (2010).

6. Clutton-Brock, T.H., Huchard, E. Social competition and selection in males and females. *Philosophical Transactions Royal Society London B Biological Sciences*, 368, 20130074 (2013).

7. Goldman, J.G. Can this sneaky chimp read minds? *BBC Future* (25 November 2013).

16 Consciousness and the Unconsciouss

1. Richards, R.J. The Impact of German Romanticism on Biology in the Nineteenth Century. *University of Chicago*, https://philpapers. org/rec/RICTIO-7 (2011).

2. Hendrix, J.S. *Unconscious Thought in Philosophy and Psychoanalysis*, p. 19 (Palgrave Macmillan, 2015).

3. McLeod, S. Id, Ego and Superego. *Simply Psychology* (3 April 2023).

4. Hughes, B. Genius of the Modern World. *BBC Four*, https://www. bbc.co.uk/programmes/b07gpjmm (June 2016).

5. LeFanu, J. Wrong image of Freud has entered the subconscious. *The Telegraph* (14 May 2006).

6. de Guerrero, M.C.M. Going covert: Inner and private speech in language learning. *Language Teaching*, 51, 1–35 (2018).

7. Meares, M. *The Metaphor of Play: Origin and Breakdown of Personal Being*, p. 38 (Routledge, 2002).

8. Quinodoz, J.M. *Reading Freud: A Chronological Exploration of Freud's Writings*, p. 145 (Routledge, 2004).

9. Young, R. Back to Bakhtin. *Cultural Critique*, 2, 71–92 (1985).

10. Saldanha, D. Antonio Gramsci and the Analysis of Class Consciousness: Some Methodological Considerations. *Economic and Political Weekly*, 23, 11–18 (1988).

11. Suzman, J. Why 'Bushman banter' was crucial to hunter-gatherers' evolutionary success. *The Guardian* (29 October 2017).

12. Suzman, J. How Neolithic farming sowed the seeds of modern inequality 10,000 years ago. *The Guardian* (5 December 2017).

13. Cunliffe, B. Against the Grain by James C Scott review – the beginning of elites, tax, slavery. *The Guardian* (25 November 2017).

14. Aristotle. Politics. *MIT*, http://classics.mit.edu/Aristotle/ politics.1.one.html (2009).

15. Hayward, L. Slavery in Ancient Rome: The Journey to Freedom. *The Collector* (13 September 2021).
16. Hodal, K. One in 200 people is a slave. Why? *The Guardian* (25 February 2019).
17. Cox, R. Inequality gap: Growing gulf between rich and poor leaves 42 people with same wealth as world's 3.7bn worst off. *The Independent* (22 January 2018).

17 Modernity and its Contradictions

1. Lindberg, I. Dictatorships advancing globally. *University of Gothenburg*, https://www.gu.se/en/news/dictatorships-advancing-globally (3 March 2022).
2. Segal, E. New Surveys Show Burnout Is An International Crisis. *Forbes* (15 October 2022).
3. Fonseca, G.L. Guide to Adam Smith's Wealth of Nations. *The History of Economic Thought* (2022).
4. Eschner, K. In 1913, Henry Ford Introduced the Assembly Line: His Workers Hated It. *Smithsonian Magazine* (1 December 2016).
5. Ghosh, S. Peeing in trash cans, constant surveillance, and asthma attacks on the job: Amazon workers tell us their warehouse horror stories. *Business Insider* (5 May 2018).
6. Rasmussen, D.C. Does 'Bettering Our Condition' Really Make Us Better Off? Adam Smith on Progress and Happiness. *The American Political Science Review*, 100, 309–318 (2006).
7. Elster, J. *Karl Marx: A Reader*, p. 39 (Cambridge University Press, 2008).
8. Ripstein, A. Commodity Fetishism. *Canadian Journal of Philosophy*, 17, 733–748 (1987).

18 Sanity and Madness

1. Kirk, A., Scott, P., Wilson, J. World Mental Health Day: The charts that show that the UK is in the midst of a mental health awakening. *The Telegraph* (14 May 2017).
2. Balter, M. Schizophrenia's Unyielding Mysteries. *Scientific American*, 316, 54–61 (1 May 2017).
3. Devlin, H. Radical new approach to schizophrenia treatment begins trial. *The Guardian* (3 November 2017).
4. Gibney, P. The Double Bind Theory: Still Crazy-Making After All These Years. *Psychotherapy in Australia*, 12, 48–55 (2006).

5. Kozulin, A. *Vygotsky's Psychology: A biography of ideas*, pp. 226–8 (Harvester Wheatsheaf, 1990).

6. Rund, B. Attention, Communication, and Schizophrenia. *The Yale Journal of Biology and Medicine*, 58, 265–273 (1985).

7. Smyth, C. Drugs remain an enduring medical mystery. *The Times* (21 July 2018).

8. Boseley, S. Antidepressants: is there a better way to quit them? *The Guardian* (22 April (2019).

9. Swanson, J. Unraveling the Mystery of How Antidepression Drugs Work. *Scientific American* (10 December 2013).

10. Albert, P.R. Adult neuroplasticity: A new 'cure' for major depression? *Journal of Psychiatry and Neuroscience*, 44, 147–150 (2019).

11. Knapton, S. Depression is a physical illness which could be treated with anti-inflammatory drugs, scientists suggest. *The Telegraph* (8 September 2017).

12. MacGill, M. What is depression and what can I do about it? *Medical News Today* (5 August 2022).

13. Ledford, H. First robust genetic links to depression emerge. *Nature*, 523, 268–269 (2015).

14. Lu, G., et al. Role and Possible Mechanisms of Sirt1 in Depression. *Oxidative Medicine and Cellular Longevity*, 2018, 8596903 (2018).

15. Brown, G.W., Harris, T. *Social Origins of Depression: A Study of Psychiatric Disorder in Women* (Free Press, 1978).

16. Oatley, K. Depression: crisis without alternatives. *New Scientist* (June 1984, pp. 29–31).

17. Beck, J. The Running Conversation in Your Head. *The Atlantic* (23 November 2016).

18. Nordqvist, C. Understanding the symptoms of schizophrenia. *Medical News Today* (6 January 2023).

19. Sass, L.A., Parnas, J. Schizophrenia, consciousness, and the self. *Schizophrenia Bulletin*, 29, 427–444 (2003).

19 How Ideas Change

1. Edgerton, W.F. The Strikes in Ramses III's Twenty-Ninth Year. *Journal of Near Eastern Studies*, 10, 137–145 (1951).

2. Mark, J.J. The First Labor Strike in History, *World History Encyclopedia* (4 July 2017).

3. Coldrick, M. Margaret Thatcher and the pit strike in Yorkshire. *BBC News* (8 April 2013).

4. Townsend, M. Miners' strike: 'All I want is for someone to say: I'm sorry'. *The Guardian* (1 December 2012).

5. Gillan, A. 'I was always told I was thick. The strike taught me I wasn't'. *The Guardian* (10 May 2004).

6. Hayes, D. Thirty years on: the Socialist Workers Party and the Great Miners' Strike. *International Socialism Journal* (2 April 2014).

7. Kellaway, K. When miners and gay activists united: the real story of the film Pride. *The Guardian* (31 August 2014).

8. Lehtonen, M. *The Cultural Analysis of Texts*, p. 157 (Sage Publications Ltd, 2000).

9. Hill, C. *The World Turned Upside Down: Radical Ideas During the English Revolution.* (Penguin, 1991).

10. Grenby, M.O. Writing Revolution: British Literature and the French Revolution Crisis, a Review of Recent Scholarship. *Literature Compass* 3, 1351–1385 (2006).

11. Hartley, C.A., Somerville, L.H. The neuroscience of adolescent decision-making. *Current Opinion in Behavioral Sciences*, 5, 108–115 (2015).

12. Marx, K. *The Eighteenth Brumaire of Louis Bonaparte*, https://www.marxists.org/archive/marx/works/1852/18th-brumaire/ch01.htm (1852).

13. Tiruneh, G. Social Revolutions: Their Causes, Patterns, and Phases. *SAGE Open*, 4, 3, (2014).

14. Hunt, T. Britain's very own Taliban. *New Statesman* (17 December 2010).

15. Benn, T. The Levellers and the Tradition of Dissent, *BBC History* (17 February 2011).

16. Harris, C. Russia's February Revolution Was Led by Women on the March. *Smithsonian Magazine* (17 February 2017).

17. Vallance, T. Fresh Perspectives on the Levellers. *History Today* (6 June 2017).

18. Weissman, S. The Golden Era. *Jacobin Magazine* (18 December 2017).

19. Luskin, B.J. MRIs Reveal Unconscious Bias in the Brain. *Psychology Today* (7 April 2016).

20. Rouault, M., Drugowitsch, J., and Koechlin, E. Prefrontal mechanisms combining rewards and beliefs in human decision-making. *Nature Communications*, 10, 301 (2019).

21. Bond, M. Mob Mentality. *Slate* (17 March 2015).

22. Trew, B. How distaste of LGBT people in Egypt has turned into state-sponsored persecution, *The Independent* (17 May 2015).

23. Knapton, S. Nine in 10 people would electrocute others if ordered, rerun of infamous Milgram Experiment shows. *The Telegraph* (14 March 2017).

20 Future of Mind

1. Birrell, I. Greta Thunberg teaches us about autism as much as climate change. *The Guardian* (23 April 2019).

INDEX

ARTIFICIAL INTELLIGENCE

Modern Magic or Dangerous Future?

YORICK WILKS

AI expert Yorick Wilks takes a journey through the history of artificial intelligence up to the present day, examining its origins, controversies and achievements, as well as looking into just how it works. He also considers the future, assessing whether these technologies could menace our way of life, but also how we are all likely to benefit from AI applications in the years to come. Entertaining, enlightening, and keenly argued, this is the essential one-stop guide to the AI debate.

ISBN 9781785785160 (paperback) / 9781785785177 (ebook)

We inhabit a planet in peril. Our once temperate world is locked on course to become a hothouse entirely of our own making.

Hothouse Earth provides a post-COP26 perspective on the climate emergency, acknowledging that it is now impossible to keep this side of the 1.5°C climate change guardrail. The upshot is that we can no longer dodge the arrival of disastrous – and all-pervasive – climate breakdown that will come as a hammer blow to global society and economy.

Bill McGuire explains the science behind the climate crisis and presents a blunt but authentic picture of the world bequeathed to our children and grandchildren; a world already glimpsed in today's blistering heatwaves, calamitous wildfires and ruinous floods and droughts. This picture is one we must all face up to, if only to spur genuine action to stop a harrowing future becoming a truly cataclysmic one.

ISBN 978-178578-920-5 (paperback)
ISBN 978-178578-921-2 (eBook)